Biology 102

Lecture Guided Workbook and Access Code for Online Learning Modules

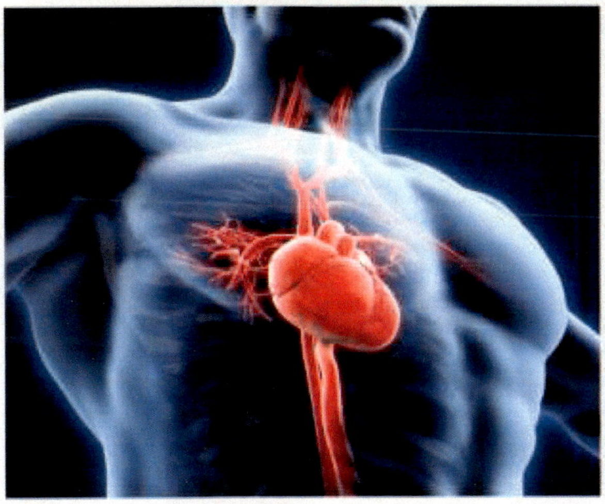

S. Mlotshwa

Access Code Registration

Biology 102
"Biological Principles II"

Step 1: Log into blackboard and select your course
Step 2: Open the "Assignments" page.
Step 3: Click the "QDE-Registration" link.
Step 4: Enter your unique access code shown below.

232 Bio 102

5PnRDTG5

Registration Problems?

"Access code has already been used" You only have to register once and will receive this response if you attempt to re-register. You may have purchased a used access code. Return to where you purchased the product for assistance.

"Access code invalid" Check that you entered the code exactly as it appears above.

Still need assistance? email: qdesales@qdepress.com. Include your name, phone number, school, course number and course name, access code as well as a brief description of your registration issue.

Biology 102
For Science Majors

Sizolwenski Mlotshwa

QDE PRESS

2017

Copyright © 2017 by QDE Press Inc.

All rights reserved.

Permission in writing must be obtained from the publisher before any part of this work may be reproduced or transmitted in any form or by any means, electronic or mechanical, including photocopying and recording, or by any information storage or retrieval system.

Printed in the United States of America.

ISBN: 978-1-938535-15-4

QDE Press Inc.

8828 Autumnbrooke Way

Montgomery, Al 36117

www.qdepress.com

Lecture Topics	Slide Number
1, Transformation of life on Earth	1
2, Evolution via Natural Selection	33
3, Speciation	73
4, Origin of Life on Earth	105
5, Phylogeny & Ring of Life	133
6, Origin of Plants	161
7, Origin of Seed Plants	185
8, Plant's Growth & Differentiation	205
9, Acquisition & Transport of Resources	245
10, Angiosperms	281
11, Plant Responses	313
12, Origin of Animals on Earth	345
13, Origin of invertebrates	361
14. Origin of Vertebrates	377
15, Anatomy & Physiology of Animals	409
16, Animal Nutrition	441
17, Circulatory & Respiratory System	485
18, Immune system	533
19, Study of Ecology	585
20, Ecosystem Ecology	625
21, Prokaryotes	661

1, Transformation of life on Earth

Evolution by means of Natural selection

- **Evolution** is the process of change that has transformed life on earth, from a simple to a very complex form.
- Evolution is also defined as **"descent with modification"**, & is known to happen via Natural selection.
- **Natural selection** is nature's way of selecting the "best adapted organisms" in a population, & then promoting them to the next generation's gene pool.
- The "best **adapted** organisms" are those that tend to survive & reproduce more offspring's than others in a population.

- Individuals with traits that are better adapted to their environment tend to produce more offspring's than those who are not better adapted
- This differential success in reproduction results in certain alleles being passed to the next generation in greater proportions and leads to the accumulation of certain traits over generations.
- These traits happen to be those that are advantageous or favorable in the current environmental conditions and further confer the species that have them, a better chance at survival & success.

- The nature's way of selecting the better adapted individuals over others and promoting their genes to the next generation in more frequency by means of more successful viable offspring's is called Natural Selection
- When enough of such traits are accumulated in the gene pool over time, it may lead to the origin of a new species, provided that the current species doesn't go extinct before it evolves into a new species

Charles's Darwin

- In 1831, Charles Darwin embarked on a five-year survey voyage around the world on the HMS *Beagle* and collected specimens of South American plants and animals

- He observed adaptations of plants and animals that inhabited many diverse environments

- His studies of specimens around the globe led him to formulate his theory of evolution and on the process of natural selection which he later published in the book *On the Origin of Species.*

- Darwin, noticed similarities among species all over the globe, along with variations, leading him to believe that they had gradually evolved from common ancestors.

- He came to believe that species survived through a process called "natural selection," where the species that successfully adapted to meet the changing environments thrived, while those that failed to evolve and reproduce died off

Adaptation in Darwin's view

- Adaptation to the environment and the origin of new species were seen as two closely related processes by Charles Darwin.

- An example of such adaptation and origin of new species is very evident in Galápagos finches

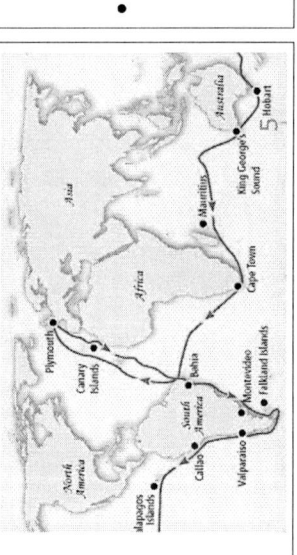

Artificial selection

- **Artificial Selection** is defined as the choosing of desirable traits by an outside source other than the organism itself or natural selection

- Darwin noted that humans have modified other species by selecting and breeding individuals with desired traits, a process called **artificial selection**

- Charles Darwin made use of artificial selection to help gather evidence for his proposed mechanism for how evolution occurs

- After studying finches, Darwin was able to show that he could choose which traits were desirable in pigeons and increase the chances for those to be passed down to the offspring by breeding two pigeons with the trait.

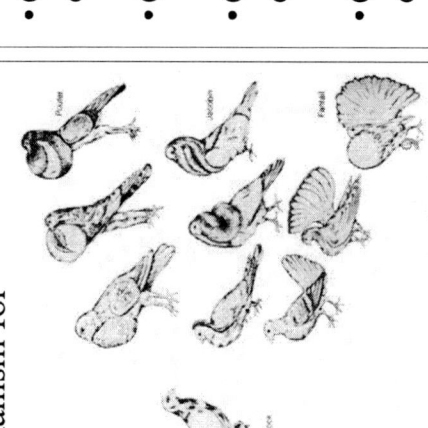

Darwin's Observations

- Observation #1: Members of a population of the same species often vary greatly in their traits

- Observation #2: Traits are inherited from parents to offspring

- Observation #3: All species are capable of producing more offspring than the environment can support

- Observation #4: Due to lack of food or other resources, many of the offspring do not survive

Darwin's Inferences

- Inference #1: Individuals whose inherited traits give them a better chance at surviving and reproducing tend to leave more offspring than other individuals

- Inference #2: This unequal ability of individuals to survive and reproduce will lead to the accumulation of certain advantageous traits in the population over generations

- Darwin Inferences helped to develop two main ideas:
  1. Evolution explains life's unity and diversity and
  2. Natural selection is a cause of this adaptive evolution

Further understanding Darwin's inferences

- If some heritable traits happen to be advantageous (traits that increase an organisms chances of survival), then these traits end up accumulating in populations over generations.

- Such <u>accumulations</u> of traits over generations, further increases the frequency of the individuals with those traits.

- This happens because individuals with those advantageous traits have a better chance at survival and so they end up <u>making more offspring</u> than those who didn't have those traits.

Unpredictable changes in the environment

- If due to some sudden changes the environment turns hostile, then individuals who were the best adapted to previous environments, may not be well adapted to this new environment

- Moreover, those who were not the best fit before, may end up emerging as a stronger species.

- In conclusion, the genetically determined characteristics or traits that we inherit or that are introduced in us via mutations, may or may not give us an advantage in times of crisis

- In some cases the variations of genetic code that we inherit or mutations that we undergo are neutral, sometime they are advantageous whereas sometimes they are outright deadly.

- How these inherited traits play out, whether to the success of a species or to it's demise, can only be known with time.

- What needs to be taken away is that the traits that are beneficial to us today, the ones that makes us best adapted to the our current environments, could tomorrow be responsible for our peril in the wake of a different environment.

Scientific evidence to support Evolution

1. The Fossil evidence

- Fossils are remains/traces of organisms from the past, found in layers or strata in the sedimentary rock

- The fossil record provides evidence of changes within groups over time, it also shows origin of new species, and the extinction of others.

- Paleontologists have discovered fossils of many such transitional forms

Fossil evidence

- Since the 1920's, there have been hundreds of well-dated intermediate fossils found in Africa.

- These were transitional species leading from apes to humans over the last 6-7 million years.

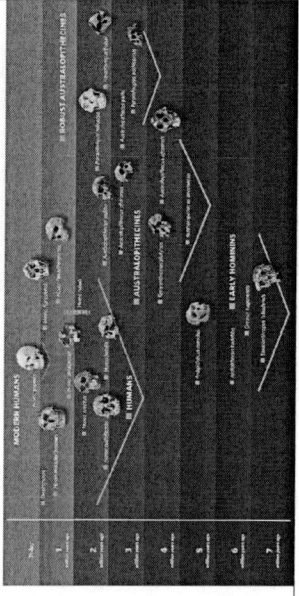

Fossil Record

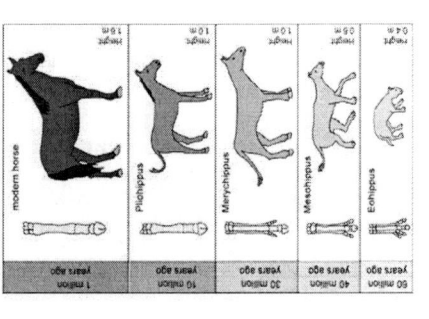

One of the few animals for which we have a fairly complete evolutionary record is the horse because all the main stages of the evolution of the horse have been preserved in fossil form.
Over 60 million years, **the horse evolved from a dog-sized creature** that lived in rainforests into an animal adapted to living on the plains and standing up to 2 meters high.

2. Homology

- **Homology** is similarity resulting from common ancestry
- **Homologous structures** are anatomical resemblances that represent variations on a structural theme present in a common ancestor
- Comparative embryology reveals anatomical homologies not visible in adult organisms

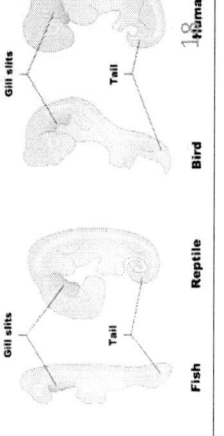

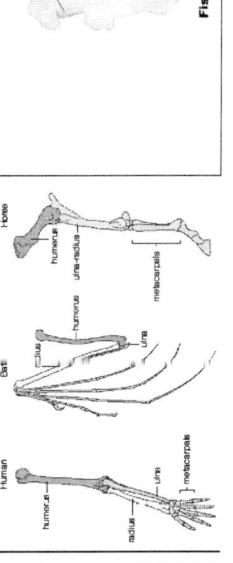

Understanding Homologies via evolutionary trees

- An **evolutionary tree** can be used to explain homologies
- Evolutionary trees are hypotheses about the relationships among different groups
- Evolutionary trees can be made using different types of data, for example, anatomical and DNA sequence data
- A simple analogy to evolutional tree →

3. *Chemical & Molecular Unity of life*

- All living things on earth share the ability to create complex molecules out of carbon and a few other elements.
- 99% of the proteins, carbohydrates, fats, and other molecules of living things are made from only 6 of the 92 occurring elements.
- All of the tens of thousands of types of proteins in living things are made of only 20 kinds of amino acids.
- Despite the great diversity of life, the simple language of the DNA code (4 bases- ATGC) is the same for all living things.

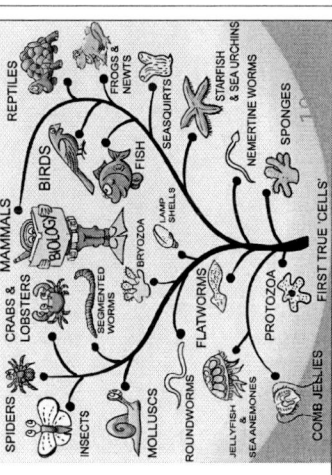

4. Geographic Distribution of Related Species

- Geographic Distribution of Species (**Biogeography**) formed an important part of Darwin's theory of evolution

- Earth's continents were formerly united as one big land mass, "Pangea" but have since separated by **continental drift** due to plate tectonics.

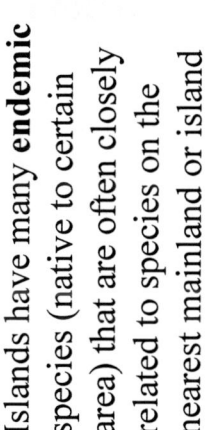

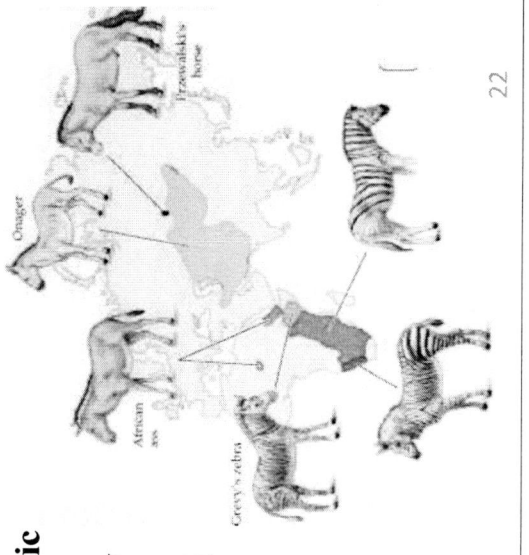

- Islands have many endemic species (native to certain area) that are often closely related to species on the nearest mainland or island

- Major isolated land areas and island groups often evolved their own distinct plant and animal communities.

- An understanding of continental drift and the current distribution of species allows us to predict when and where different groups evolved in their timeline

- For example: Darwin observed that mammals were almost never naturally present on islands that were hundreds of miles away from the nearest mainland.

- He concluded that it was almost impossible for large terrestrial animals to get to isolated islands that were so far off from mainland, as it required traveling over large masses of water & thus they never had a chance to evolve in those environments.

5. Bacterial resistance to antibiotics

- Bacteria colonies build up resistance to antibiotics through evolution.

- Since bacteria have a very small generation time, the odds of a mutation happening in a colony of bacteria during DNA replication is much higher.

- As a result, when an antibiotic is applied, the initial dose kills most bacteria, leaving behind only those few bacteria's that happen to have the mutations necessary to resist the antibiotics.

- In subsequent generations, the resistant bacteria reproduce, forming a new colony where every member is immune because they have inherited immunity from the survivors.

- This is natural selection in action. The antibiotic is "selecting" for organisms which are resistant, and killing any that are not.

➢ An example of mosquito's with and without resistant alleles

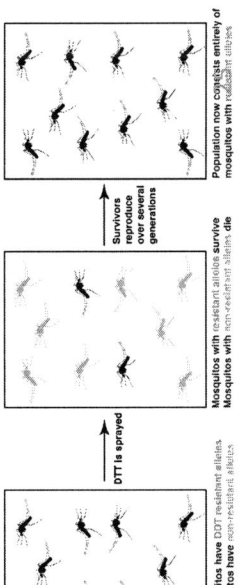

Convergent evolution

- **Convergent evolution** is the process whereby organisms that are not closely related, independently **evolve** similar or analogous traits

- The two species in Convergent evolution are not related via a recent common ancestor

- But instead due to an adaption they undergo in a similar environment, found elsewhere

Sugar glider and the Flying squirrel are very similar with their small rodent-like body structure

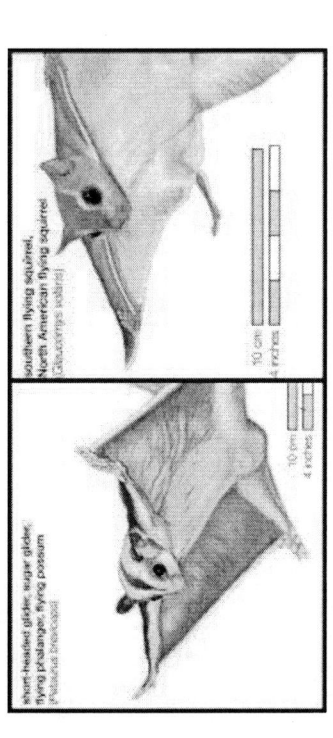

- The thin membrane that connects their forelimbs to their hind limbs is used to glide through the air in the same fashion.

Shark and Dolphin

- Another example of convergent evolution is the overall body structure of the shark and the dolphin.

- A shark is a fish & a dolphin a mammal. However, their body shape and how they move through the ocean is very similar.

- This is not because they are related via a recent common ancestor, but because they live in similar environments and adapted in similar ways in order to survive in those environments.

- Convergent evolution leads to similarity between organisms but does not provide information about ancestry like the Divergent evolution.

- Divergent evolution (or better known as just "evolution") is one where similarity arises due to common ancestry.

- Similar or analogous traits via convergent evolution arise when groups independently adapt to similar environments in similar way

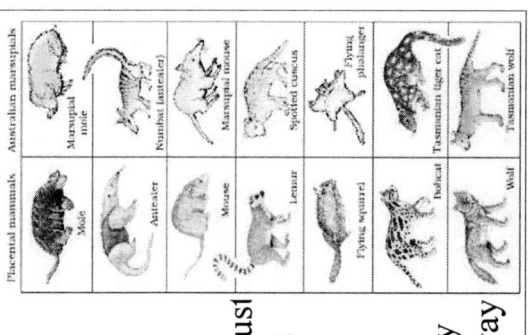

2, Evolution via Natural Selection

Mutation & Sexual reproduction produce the genetic variation that make evolution possible

- Two processes, mutation and sexual reproduction, produce the variation in gene pools that contributes to differences among individuals
- Gene pool of a population consists of all the alleles for all the loci in all the individuals of that population
- Variation in individual genotype(genetic makeup) leads to variation in individual phenotype (physical appearance) but not all phenotypic variation is heritable

- **Genetic variations** in populations are the working grounds for Natural selection & the Natural selection can only act & select the variations that are phenotypically present.
- If a genetic variation(genotype) doesn't show up in the phenotype of the organism then it can not be acted upon.
- If it can't be acted upon, then that organism neither has an advantage nor a disadvantage for having that variation in his genotype.
- Hence, such a variation would have no contribution towards the evolution of the animal.

a) *Genetic Variation Within a Population*

- Exists at 2 levels, as Gene variability & Nucleotide variability

1. Gene variability can be quantified as **Average heterozygosity** which measures average percent of loci that are heterozygous in a population (which means different alleles at a given locus)

e.g.- Drosophila Melanogaster has about 13,700 genes in its genome. On average a fruit fly is heterozygous for about 1,920 of its loci(14%) and homozygous for the remaining genes.

We can therefore say that a Drosophila melanogaster population has an average heterozygosity of 14%.

2. Nucleotide variability is measured by comparing the DNA sequences of two individuals in a population and then averaging the data from many such comparisons.

e.g.- The genome of D. Melanogaster has about 180 million nucleotides and the sequences of any two flies differ on average by approximately 1.8 million(1%) of their nucleotides.

We can therefore say that a Drosophila melanogaster population has a nucleotide variability of 1%.

---

b) *Variation Between Populations*

- Variations between populations exist due to Geographic variations.

- When populations are isolated geographically and are unable to mate with one another, they end up evolving independent of one another.

- **Effects of Random Factors:** Factors besides Natural selection (such as Mutations) when introduced in individuals of a certain population end up remaining in that population & remain missing in other populations due to geographic isolation

---

Genetic Variation
Within a Population

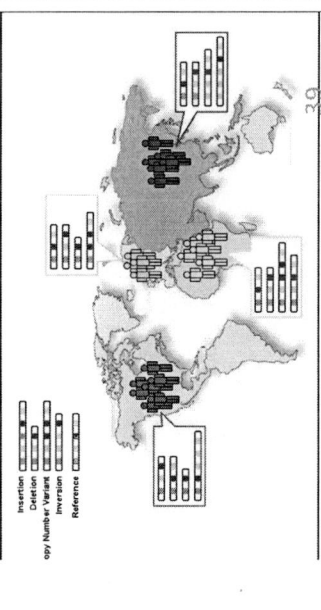

Genetic Variation
Between Populations

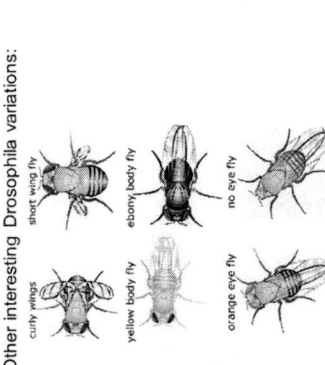

Mutations

Mutations are changes in the nucleotide sequence of DNA and are the ultimate source of new alleles in a population

- Only mutations in cells that produce gametes can be passed to offspring.

- Since most of the mutations that occur take place in somatic cells, they are lost when the individual dies.

- Mutation rates are low in animals and plants but higher in viruses due to the short generation time.

1. Point Mutations - a change of one base of a gene, the effects of point mutations can vary:

Neutral: Mutations in noncoding regions of DNA are often harmless and moreover, the Mutations in a gene might not affect protein production because of redundancy in the genetic code

Harmful: Mutations that result in a change in protein production are often harmful

Positive: Mutations can sometimes increase the fit between organism and environment

2. Chromosomal Mutations

- Chromosomal Mutations alter gene number and sequence

- Mutations leading to deletions, insertions, translocations and duplications of Chromosomes are mostly harmful

- Duplicated genes sometimes take over new functions by additional mutations

Sexual Reproduction

- Sexual reproduction shuffles existing alleles into new combinations

- In organisms that undergo sexual reproduction, the recombination of alleles is a more significant impetus than mutations to produce the genetic differences that lead to evolution

Sexual Reproduction

- Shuffling of alleles happens due to crossing over, independent assortment and random fertilization

- The combined effects of these mechanisms ensure that with each new generation the new combination of alleles provides for the genetic variation that makes evolution possible.

Natural selection, genetic drift, and gene flow act upon a population to alter the allele frequencies

The 3 major factors responsible for the most evolutionary change by altering the allele frequencies are:

1. Natural selection
2. Genetic drift
3. Gene flow

1. Natural Selection

- Individuals in a population show variations in their heritable traits.
- Individuals with traits that are better adapted to their environment tend to produce more offspring than those who are not better adapted
- This differential success in reproduction results in certain alleles being passed to the next generation in greater proportions and leads to the accumulation of certain traits over generations.

2. Genetic Drift

- The process in which chance events cause allele frequencies to fluctuate unpredictably from one generation to the next
- The smaller a sample, the greater the chance of deviation from a predicted result
- The 2 examples of genetic drift having a significant impact on population are
   1. **Founder effect**
   2. **Bottleneck effect**

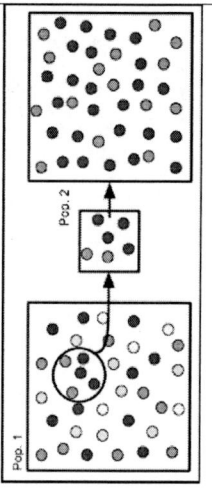

1. The **founder effect** occurs when a few individuals become isolated from a larger population

- A founder effect occurs when a new colony is started by a few individuals of the original population. This small population size may have:
  a) reduced genetic variation from the original population

  or

  b) a non-random sample of the genes in the original population.

2. The **bottleneck effect** is a sudden reduction in population size due to a change in the environment

- The resulting gene pool may no longer be reflective of the original population's gene pool

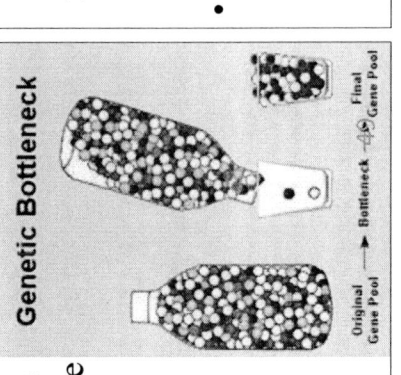

- When that happens the new population may not be able to adapt to new conditions, such as climatic change or the predator prey interactions because of being outnumbered and less diverse

3. Gene Flow

- **Gene flow** consists of the transfer of alleles among populations by movement of fertile individuals from one region to another.

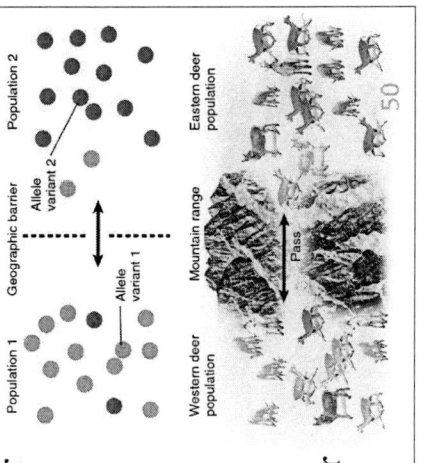

- Gene flow tends to reduce differences between populations over time & is more likely to alter allele frequencies than mutation

Gene flow and Fitness of a population

- Depending on the genes involved, gene flow can decrease or increase the fitness of a population

- An individual's **fitness** includes its ability to survive, find a mate, produce offspring and eventually pass on the genes to the next generation's gene pool.

- Also, we need to keep in mind that the fittest individuals are not necessarily the strongest, fastest, or biggest in the populations.

- Fittest individuals are simply those that are best adapted to their environment and have the ability to make viable, healthy and fertile offspring's

- Moreover, Relative Fitness is defined as the contribution an individual makes to the gene pool of the next generation, relative to the contributions of other individuals

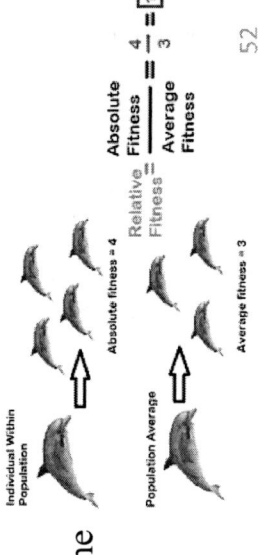

$$\text{Relative Fitness} = \frac{\text{Absolute Fitness}}{\text{Average Fitness}} = \frac{4}{3} = \boxed{1.33}$$

Gene flow decreasing the fitness of a population

- In certain grass, near the copper mines, the alleles for copper tolerance are beneficial in populations but are harmful to populations away from the copper mines
- The pollen with the help of wind and insects moves from one place to another and the alleles for copper tolerance are hence introduced in populations where they aren't welcomed
- This movement of unfavorable alleles into a population results in a decrease in fit between organism and environment

Gene flow increasing the fitness of a population

- An example of Gene flow increasing the fitness of a population is the spread of insecticide resistant gene in mosquitoes.
- Insecticides have been used to target mosquitoes that are a carrier of West Nile virus that cause malaria
- Over a short period of time these insecticide resistant alleles which existed only in a few individuals of small populations, have now spread over to new populations by gene flow, thereby increasing the fitness of the new population

Mechanism that consistently causes adaptive evolution?

- Only natural selection consistently results in adaptive evolution
- Because the environment can change, adaptive evolution is a continuous process
- Genetic drift and gene flow do not consistently lead to adaptive evolution as they can both increase or decrease the match between an organism and its environment
- When there is such a decrease in the match, it doesn't promote evolution

- Natural selection increases the frequencies of alleles that enhance survival and reproduction
- Natural Selection favors certain genotypes by acting on the phenotypes of organisms. It does that using 1 of the 3 modes

1. Directional selection favors individuals at one end of the phenotypic range

2. Disruptive selection favors individuals at both extremes of the phenotypic range

3. Stabilizing selection favors intermediate variants

Sexual Selection- Natural selection for mating success

- **Sexual selection** is a form natural selection in which individuals with certain heritable characteristics are more likely than other individuals to obtain mates

- Sexual selection makes many organisms go to extreme lengths for sex: peacocks maintain elaborate tails, elephant seals fight over territories, and fruit flies perform dances

- Sexual selection can be either Intrasexual or Intersexual in nature

Sexual Dimorphism- a consequence of Sexual Selection

- Which ever be the type, sexual selection can result in **sexual dimorphism**, marked differences between the sexes in secondary sexual characteristics

Directional, Disruptive, and Stabilizing Selection

1. **Intrasexual selection** is competition among individuals of one sex (often males) for mates of the opposite sex

2. **Intersexual selection**, often called mate choice, occurs when individuals of one sex (usually females) are choosy in selecting their mates

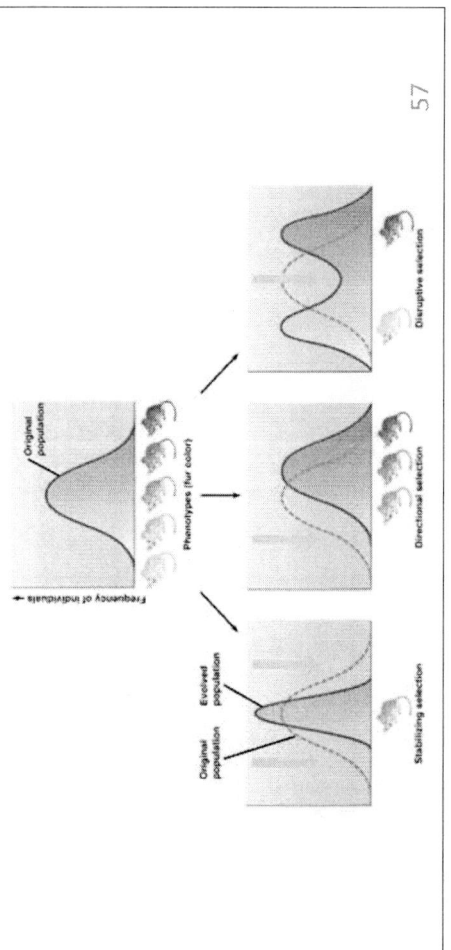

The Preservation of Genetic Variation

- Natural selection constantly works towards promoting the better adapted individuals to the next generation but while doing so, it subsequently reduces the genetic variation over generations

- It does so because the individuals with less fitness end up making lesser offsprings and if this process goes on unchallenged for too long, Natural selection would entirely let go of such individuals with lower fitness and subsequently eliminate the very genetic variation that made Natural selection possible in the first place.

- Sexual selection is powerful enough to produce features that could be harmful to the individual's survival.

- For example, extravagant and colorful feathers or fins are likely to attract members of the opposite sex but at the same time attract the predators as well

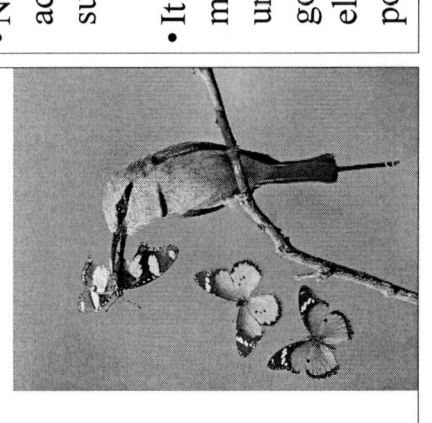

- Wing color patterns of butterflies must perform different signaling functions to avoid predators and attract potential mates. When they fail to do so the results can be life threatening

- In other words, Natural selection if not countered will eliminate the very cause of its existence (Genetic variation) by playing a role in leaving behind all those that are unfit (including those that may become a better fit in a few generations down the line)

- Hence, some processes must exist in nature that help restore and preserve the Genetic Variation that Natural selection, Genetic drift and Gene flow significantly curtail and narrow down by acting upon it

Various mechanisms that help to preserve genetic variation in a population

1. **Diploidy**: maintains genetic variation in the form of hidden recessive alleles

2. **Balancing selection**: There are 2 types of balancing selection
   a) Heterozygote advantage
   b) Frequency-dependent selection

3. **Neutral Variation**

2. (a) Heterozygote advantage occurs when heterozygotes have a higher fitness than do both homozygotes

- E.g.: The sickle-cell allele causes mutations in hemoglobin and causes sickle-cell anemia if it occurs in a homozygous pair

but if it occurs in a heterozygous combination (1 sickle & 1 normal allele), it offers malaria resistance while keeping the individual safe from developing sickle cell anemia

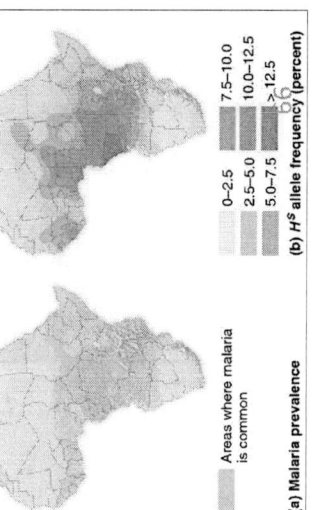

The sickle-cell allele

- The sickle cell disease has a recessive pattern of inheritance: and only individuals with two copies of the sickle-cell allele have the disease.

- People with just one copy are healthy and thus sickle-cell allele is maintained in human populations as a Heterozygote advantage, in areas where malaria is prevalent.

Sickle-cell allele & Heterozygote advantage

- The Protozoa "Plasmodium" (carried in the gut of mosquito) prefers individuals with round RBC's and not those with sickle cell, since round RBC's make better host for the eggs laid in the host body for the spread of the plasmodium parasite

- Having only one of the copy of the "s" allele makes some of the RBC's sickle in shape, just enough to reduce the chances of the "Plasmodium" to lay eggs on such a host, and consequently offering a good resistance to the malarial parasite

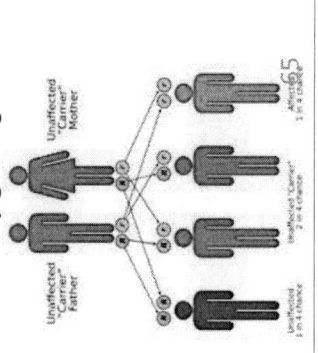

2 (b) Frequency-dependent selection, is the hypothesis that as a phenotype becomes more common, it becomes more vulnerable to a predator

- As a result, Natural selection will favor whichever phenotype is less common in a population

- As in the example with African cichlid (small fish in the figure), which exists in 2 kinds, left mouth vs right mouth where Natural selection as we would see is frequency dependent

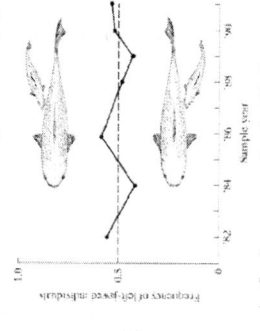

African cichlid

- When the left mouth fish (as in top figure) attacks the prey fish (the bigger fish) from the right side in higher frequency the prey fishes are on a lookout at their right sides while not carefully guarding their left.

- When that happens it provides an advantage to the left mouth fishes as they get to attack more of the prey fish and thereby becomingg a better fit in that new changed environment

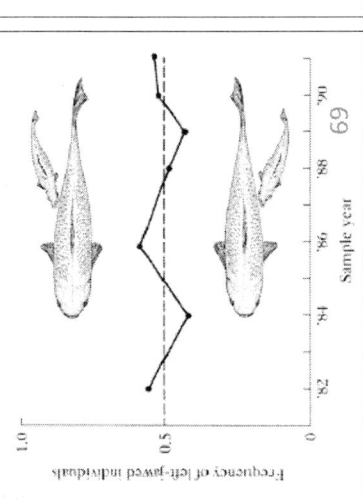

- But soon enough as the attacks on the prey fish increase from the left due to high frequency of left mouth predator fishes, the prey fishes become more aware and starts guarding their left side and not so much of the right

- This turn of events will now favor the right mouth fish to again increase its frequency over the left mouth fish and the process repeats itself multiple times

- As we learn in this example Natural Selection constantly favors the phenotype that is less common in the population to makes sure that nature doesn't loose out on the genetic variation

c) Neutral Variation

- Some of the Genetic variation that is not acted upon by Natural selection is **Neutral variation**.

- Natural selection is indifferent to Neutral variation, since most of it doesn't exist in the phenotypic expression of the organism.

- It does so mainly by existing in the **noncoding regions** of the DNA & hence remaining only as a Genotype & not a Phenotype that can be influenced by being acted upon by Natural selection.

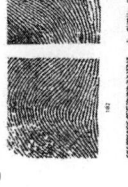 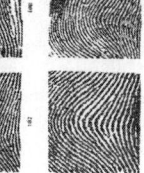

- In cases where it **does exist as a phenotypes** (fingerprints or eye color) and not in the non coding regions, it still doesn't confer an advantage or a disadvantage to the individual

- Such Neutral variation offers no benefit or harm to individuals but rather exist as a hidden reservoir of variation which may express itself when least expected and most required.

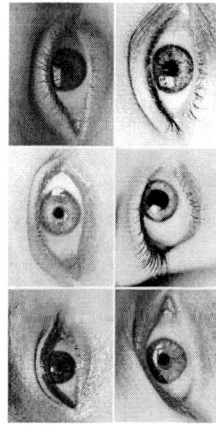

3. Speciation

Speciation: The origin of a new species

- The process by which one species splits into two or more species is called **Speciation**.

- For speciation to take place, we need a reproductive barrier to interrupt the gene flow between the existing species (singular)

- If a barrier prevents the 2 populations (currently same species) from mating and exchanging genes, then such a barrier would be called a reproductive barrier.

- Such a reproductive barrier puts a population of a species in reproductive isolation with the other population of the same species

- Over a considerable period of time (depending on the species) if the gene flow remains interrupted, it may lead to the origin of a new species, as shown below

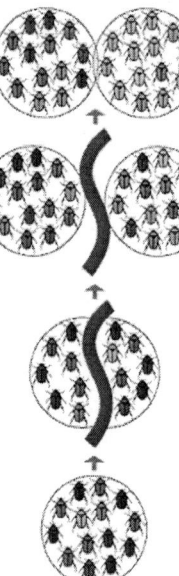

- This would happen because if the 2 populations happen to be in different/ isolated regions and will then independently respond to the factors like Natural Selection, Genetic Drift and Gene Flow when these act upon them.

- Moreover, these factors may significantly act upon the genetic variation of only 1 population and almost minimally towards the other.

- This process can be further intensified if an event like founder effect happens prior to the establishment of a reproductive barrier.

- Since, a highly probable consequence of founder effect is for the isolated population to be unlike the original population in the genetic make up.

- Such an event would increase the odds of Speciation many folds when acted upon by Natural selection

- This understanding of founder effect was first observed by Darwin in context of the plants and animals that he discovered on Galápagos Islands and noted to have not found them any where else on Earth

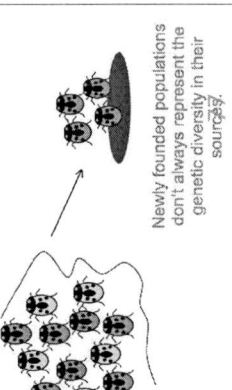

Newly founded populations don't always represent the genetic diversity in their source.

- An example of Speciation can be seen with the flightless bird below, "Cormorant" found only on the Galapagos Islands.

- Since the mainland happens to be hundreds of miles away from Galapagos, such a flightless bird could have only originated on the Galapagos island

- In the absence of land predators, the ability to fly no longer conferred an advantage for these fish-eating birds.

The biological species concept

- According to the **biological species concept** a **species** is a group of populations whose members have the potential to interbreed in nature and produce viable, fertile offspring's.

- **Reproductive isolation** is the existence of biological barriers that impede two species from producing viable, fertile offspring. There are 2 kinds of Reproductive barriers:

1. Prezygotic barriers
2. Postzygotic barriers

Pre-zygotic barriers

- **Prezygotic barriers** block fertilization from occurring by either

  – Impeding different species from attempting to mate or

  – Preventing the successful completion of mating or

  – Hindering fertilization if mating is successful

- **The 5 Types of *Pre-zygotic barriers are*-**

1. **Habitat isolation**: Two species occupy different habitats and thus encounter each other rarely, or not at all.

2. **Temporal isolation**: Species that breed at different times of the day, seasons, or different years cannot mix their gametes

3. **Behavioral isolation**: Courtship rituals and other behaviors unique to a species are effective barriers

4. **Mechanical isolation**: Morphological differences can prevent successful mating

5. **Gametic isolation**: Sperm of one species may not be able to fertilize eggs of another species

Post-zygotic barriers

- **Post-zygotic barriers** prevent the hybrid zygote from developing into a viable, fertile adult:

1. **Reduced hybrid viability**: Genes of the parent species interact in the offspring, may impair hybrid's development

2. **Reduced hybrid fertility**: Even if Hybrids development is not impaired, they may end up sterile

3. **Hybrid breakdown**: In some cases the first-generation hybrids are fertile, but when they mate, the next generation offspring's end up feeble or sterile

Pre-zygotic barriers

Post-zygotic barriers

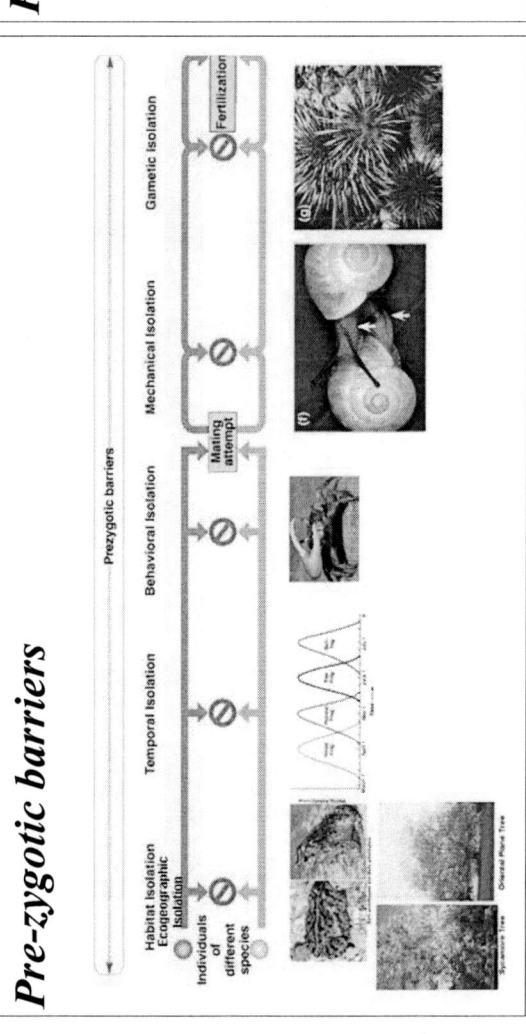

Speciation

- Speciation can take place with or without geographic separation and hence occurs in two ways:

1. **Allopatric speciation** – occurs when the 2 populations of the same species become geographically isolated from one another due to a physical barrier, eventually causing the gene flow between them to be interrupted

2. **Sympatric speciation** – speciation that occurs without the presence of a physical barrier, in such cases the gene flow is interrupted due to reasons other than a physical barrier

Allopatric speciation

- An ancestral fish population was split into two by the formation of the Isthmus of Panama about 3.5 million years ago.

- Since that time, different genetic changes have occurred in the two populations because of their geographic isolation

- These changes eventually led to the formation of different species.

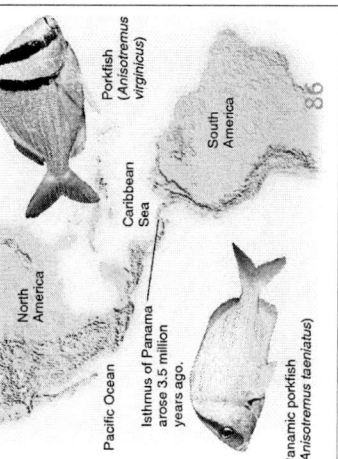

Allopatric speciation

- Reproductive isolation between populations generally increases as the distance/ barrier between them increases

- As a result, regions with many geographic barriers typically have more species than do regions with fewer barriers.

- This happens because of the interruption in the gene flow between populations, and hence different populations evolve independent of one another.

Sympatric Speciation

- In **sympatric speciation**, speciation takes place in geographically overlapping populations

- **Sympatric speciation** can occur in one of the following ways-

1. Polyploidy
2. Habitat Differentiation
3. Sexual selection

1. Polyploidy

- **Polyploidy** is a random event that leads to the presence of extra sets of chromosomes due to accidents during cell division
- After polyploidy occurs in a plant, that plant ends up being isolated from the rest, despite the absence of physical barriers.
- It happens because the other plants can now no longer endure cross pollination to make offspring's with a plant, whose chromosomal number is abnormal.
- Such cross pollination attempts would fail to create a healthy viable offspring due to the lack of genetic compatibility.
- Such plants with polyploidy hence end up being isolated from other plants and can now only successfully reproduce by either
  a) Self pollination or by
  b) Mating with plants that have undergone similar polyploidy
- The consequence of such an event leads to the creation of a subset of individuals within the main population
- Overtime, changes in the environment will affect these plants with different genotypes differently since their gene pools are no longer common and instead cutoff from one another.
- Such a subset of population which is now reproductively isolated will evolve independent of the main population.
- Polyploidy is much more common in plants than in animals
- Many important crops (oats, cotton, potatoes, tobacco, and wheat) are polyploids

2. Habitat Differentiation

- Sympatric speciation can also result from the appearance of new ecological niches
- As in the case of maggot fly where some of the flies randomly ended up on the Apple trees and chose that as their habitat vs flies who stayed on Hawthorn trees.
- The apple trees planted in America were introduced by British settlers and flies that were originally found on Hawthorn trees accidentally ended up on the Apple trees

- Now, since Apple trees provided a comfortable habitat, some flies never returned back to the Hawthorn trees and over time their gene pool remained isolated from the other flies.

- Several generations later these 2 groups of flies having fed on different foods in different habitats evolved independent of one another.

- Despite the absence of physical barrier, the flies were isolated from one another due to habitat differentiation

3. Sexual Selection

- **Sexual selection** is a form natural selection in which individuals with certain heritable characteristics are more likely than other individuals to obtain mate

- Sexual selection makes many organisms go to extreme lengths for sex: peacocks maintain elaborate tails, elephant seals fight over territories, and fruit flies perform dances

Hybrid zone and it's possible outcomes

- A **hybrid zone** is an area in which members of different species mate and produce hybrids.

- When closely related species meet in a hybrid zone, there are three possible outcomes:

  1. Strengthening of reproductive barriers
  2. Weakening of reproductive barriers
  3. Continued formation of hybrid individuals

Changes in the Hybrid Zone over Time

TIME

Reinforcement:
Hybrids are less fit than either purebred species. The species continue to diverge until hybridization can no longer occur.

Fusion:
Reproductive barriers weaken until the two species become one.

Stability:
Fit hybrids continue to be produced.

1. Reinforcement: Strengthening Reproductive Barriers

- The **reinforcement** of barriers occurs when hybrids are less fit than the parent species.

- Reinforcement of barriers is an outcome of Natural selection increasing the reproductive isolation between species

- Over time, the rate of hybridization would decrease and the species would diverge from one another with strengthening barriers until they can no longer produce a viable offspring

2. Fusion: Weakening Reproductive Barriers

- If hybrids are as fit as parents, there can be substantial gene flow between species

- If gene flow is great enough, the parent species can fuse into a single species by the weakening of reproductive barriers.

- Over time for the hybrids to persist & dominate, they would have to exploit the resources better than the parent species

- Over time Hybrids can themselves become a separate species.

3. Stability: No significant change in barriers

- During Stability of reproductive barriers, there is a continued formation of Hybrid Individuals

- Stability is the outcome when no real change takes place and where the hybrids are neither too fit nor unfit when compared to the parent species.

- In such an event, the two parent species will remain separate, but will continue to interact to produce some hybrid individuals

The rates of Speciation:

- Speciation can occur rapidly or slowly and can result from changes in few or many genes.

- The 2 kinds of Speciation based on the rate at which they take place:

1. **Gradual Pattern:** If speciation happens very slowly over several thousand generations
2. **Punctuated Equilibrium:** Speciation happening suddenly over a short period of time

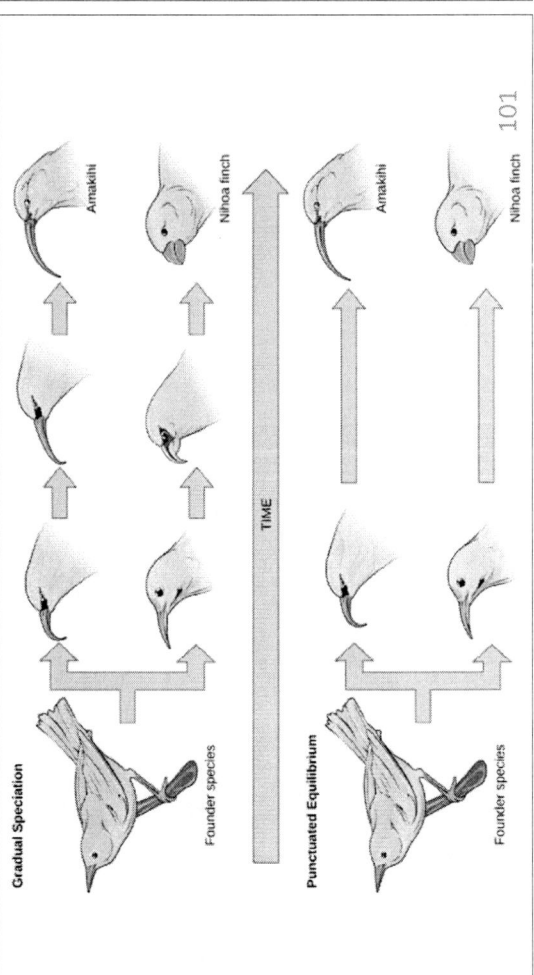

Gradual pattern vs Punctuated Equilibrium

- Depending on the species in question, speciation might require the change of only a single allele or many alleles

- The interval between speciation events can range from 4,000 years (some cichlids) to 40,000,000 years (some beetles)

- The theory of gradual pattern or gradualism rationales for a slow, constant and consistent change over a long period of time without the sudden development of a new species from one generation to the next

Gradual pattern vs Punctuated Equilibrium

- The punctuated pattern in the fossil record suggests that speciation can be rapid.

- Usually there is a period of very little change, and then a few rapid changes occur, often through mutations in the genes of a few individuals.

- The mutations that happen in certain bases on certain genes can sometimes be very effective in rapidly changing the affected individuals over just a few generations.

4, Origin of Life on Earth

Topic:1 History of Life on Earth

➤ **Conditions on early Earth made the origin of life possible**

a) Abiotic synthesis of small organic molecules

b) Joining of these small molecules into macromolecules

c) Packaging of molecules into "protobionts", droplet with membranes that maintained an internal chemistry different from their surroundings.

d) Origin of self-replicating molecules- DNA and RNA

Step 1. Synthesis of Organic molecules

- The early atmosphere of the planet earth is believed to have had nitrogen, carbon dioxide, methane, ammonia, hydrogen and water vapors

- Oparin and Haldane postulated that conditions on early Earth favored the synthesis of organic compounds from inorganic precursors.

- The energy required to make these organic molecules was provided by lightning and the intense UV radiation.

- It is also suggested that some of the organic compounds from which the first life on Earth arose came from the outer space.

- While others suggest that the synthesis of organic molecules was primarily due to submerged volcanoes and deep-sea vents that released the gases which were precursors of organic compounds.

- In 1953, Miller and Urey experiment successfully tested the Oparin-Haldane hypothesis by creating, in laboratory, the conditions that had been postulated for early earth's atmosphere

Miller-Urey experiment

- The Miller-Urey experiments produced a variety of amino acids and other organic molecules.

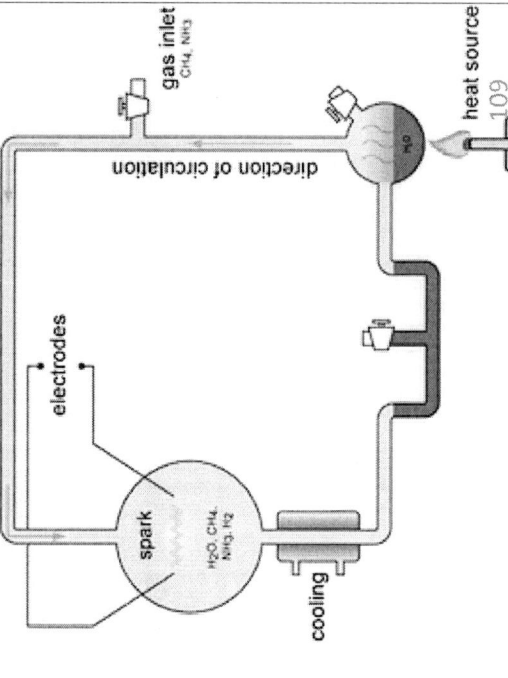

Step 2: The birth of Macromolecules

- The next step after the rise of Organic molecules was the linking of those monomers to create polymers

- With today's knowledge of enzymes and other cellular components creating polymers from monomers seems easy doesn't appear to pose a challenge

- But in the primitive atmosphere before the birth of a cell or even enzymes, the process must have taken a long time before polymers were made from monomers

- In certain experiments, researchers have been able to produce polymers, including polypeptides, after dripping solutions of monomers onto hot sand, clay, and sometimes rock.

- It happens as heat vaporizes the water in the solutions, leaving behind high concentrations of amino acids.

- Some of these concentrated amino acids then bond together and form polypeptides.

- It is postulated that dilute solutions of monomers splashing onto fresh lava in the earth's early atmosphere may have created the polypeptides and polysaccharides.

Step 3: Creation of Protobionts

- Protobionts or protocells were the precursors to the prokaryotic cells and hence to the life on planet earth.

- They were spherical structures containing some organic and inorganic compounds inside a membrane of lipid layer

- Protobionts served the basic role of maintaining a certain internal chemistry different from its outer surroundings by using its membrane as a barrier.

Step 4- Origin of RNA and DNA

- It is hypothesized that RNA was the first genetic material to develop, because unlike DNA, RNA has the ability to replicate itself. For instance:

- RNA molecules called **Ribozymes** are capable of catalyzing specific biochemical reactions, similar to the work of enzymes

- While acting as a catalyst to different reactions (as enzymes do) they also make complementary copies of their own sequence or other short pieces of RNA (as a self replicating molecule)

- Mainly because Ribozymes have demonstrated that RNA can be both genetic material and a reaction catalyst (like enzymes)

- It is believed that RNA of the primitive earth was the precursor to both DNA & RNA that we have come to know of today.

- *Some other roles of Ribozymes:*

  ➤ Ribozymes are also known to play a role within Ribosomes, as part of the large subunit ribosomal RNA, to link amino acids to one another during protein synthesis.

  ➤ They also play a role in RNA splicing, viral replication and synthesis of transfer RNA.

Topic 2: Fossil record documents history of life

- The absolute ages of fossils can be determined by **radiometric dating**

- Radiometric dating is based on decay of radioactive isotopes.

- Rate of decay is expressed by **half-life**, the time required for 50% of the parent isotope to decay. For e.g. C14 has a half life of 5370 yrs.

Radiometric dating

As the number of parent atoms decrease, the number of daughter atoms increase

- A living organism has C12 as well as C14 and when the organism dies it stops accumulating carbon, and the amount of C12 in its tissues does not change over time.

- However, C14 that the organism contains at the time of death slowly decays and becomes another element, nitrogen 14.

- Thus, by measuring the ratio of C14 to C12 in a fossil we can determine the fossil's age.

Half Life of various elements

Isotope		Half-life of parent (years)	Useful range (years)
Parent	Daughter		
Carbon 14	Nitrogen 14	5,730	100 - 30,000
Potassium 40	Argon 40	1.3 billion	100,000 - 4.5 billion
Rubidium 87	Strontium 87	47 billion	10 million - 4.5 billion
Uranium 238	Lead 206	4.5 billion	10 million - 4.6 billion
Uranium 235	Lead 207	710 million	

Topic 3 - Origin of Eukaryotes and multicellular organisms

- The origin of Eukaryotes lies in the diversification of Prokaryotes, which has been stated as theory of Endosymbiosis.

- Endosymbiosis proposes that mitochondria and plastids (chloroplasts and related organelles) were formerly small prokaryotes living within larger host cells.

- These later turned into organelles but were initially independent life forms, as a simple prokaryotic cell.

Endosymbiosis in a nutshell:

1. Start with two independent bacteria.
2. One bacterium engulfs the other.
3. One bacterium now lives inside the other.
4. Both bacteria benefit from the arrangement.
5. The internal bacteria are passed on from generation to generation.

- Mitochondria, Chloroplast and some other organelles, used to be free-living independent prokaryotic cells, until they were engulfed by a host cell that also happened to be prokaryotic.

- This amalgamation of a cell into another and their symbiotic relationship to each other lead to the birth of a Eukaryotic Cell.

- The 3 pieces of evidence that support this claim lie in the, membranes enclosing these organelles, the unique DNA, and the mode of Reproduction.

1. **Membranes**: Mitochondria and Chloroplast have their own unique membranes much like a cell's plasma membrane and are believed to have played a key role in making these early prokaryotic cells (now organelles), independent

2. **DNA**: Mitochondria & Chloroplast have their own unique DNA which falls under the category of extra nuclear DNA & is independent of the Chromosomal DNA inside the nucleus.

3. **Mode of Reproduction**: Mitochondria, like Chloroplast reproduces in a fashion very similar to binary fission of prokaryotes where the parent makes 2 identical copies

Topic 4- continental drift, mass extinctions and adaptive radiations

- **Continental drift, mass extinctions and adaptive radiations**, led to rise and fall of dominant groups and significantly shaped the history of life on earth.

1. **Continental drift-** Continents move towards or away from each other. Three major one's happened so far where land masses came together to make supercontinent (One huge landmass) and then later broke apart.

Continental drift: shaping the planet and leading to mass extinctions

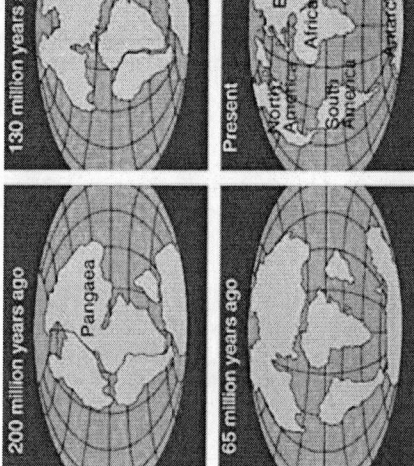

Continental drift leads to:

- New boundaries/ new masses of land which affects the habitat of the organisms, which leads to Allopatric speciation.

- Climate change- which may lead to species adapting to new environment or getting extinct.

- **Mass Extinctions**- Large numbers of species (over 50%) are wiped off from the face of the earth and become extinct forever. Has already happened 5 times in the history of planet.

Mass Extinction leads to:

- **Adaptive radiation**- these are period of evolutionary change in which group of organisms form many new species whose adaptation allow them to fill different ecological roles or niche in their communities.

- E.g. Mammals underwent adaptive radiation after extinction of Dinosaurs. Mammals increased in size and complexity by taking advantage of new environmental

Adaptive radiation of Marsupials

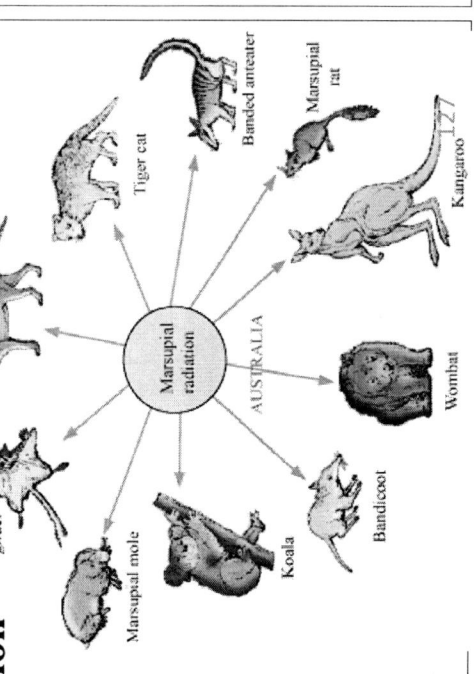

Topic 5 : Changes in sequence and regulation of gene

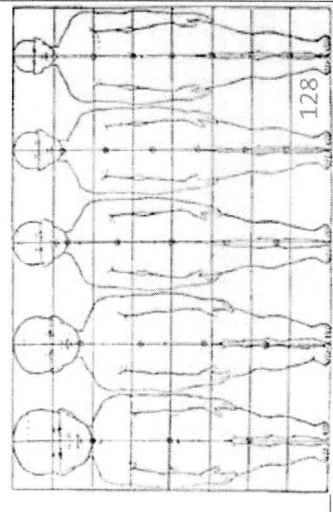

a) **Heterochrony** is an evolutionary change in the rate or timing of developmental events.

- Different growth rate of body parts is relative to each other.

a) Heterochrony can alter the timing of reproductive development which leads to a condition called paedomorphosis.

b) Paedomorphosis is the condition where the rate of reproductive development has accelerated in comparison with non reproductive organs.

- As a result a sexually mature stage of a species may retain body features that were juvenile structures in an ancestral species.

c) Homeotic genes are genes which regulate the development of anatomical structures in various organisms such as insects, mammals, and plants.

- They help determine such basic features as where wings & legs will develop on a bird or how a flower's parts are arranged.

- *Hox* genes (a class of homeotic genes) have had a huge impact in morphology of Animals

- For e.g. Duplication of Hox genes, Introduced Backbone to Invertebrates and led to evolution of Vertebrates.

Drosophila's Homeotic genes

&

Mammalian Hox genes

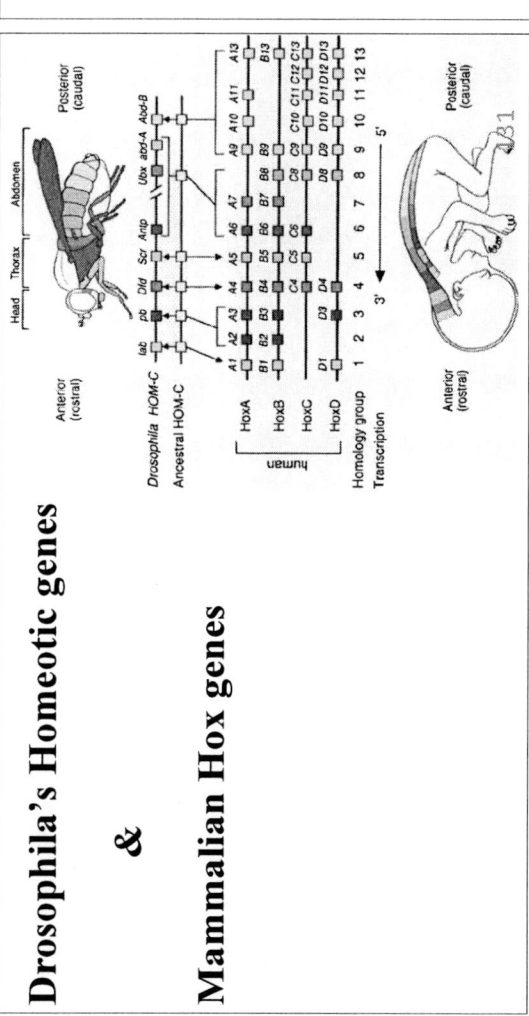

5, Phylogeny & Ring of Life

Phylogeny and tree of Life

- **Phylogeny-** is the evolutionary history of a species or group of species.
- Phylogenies depict evolutionary relationships
- Evolutionary relationships can be depicted in a branching **phylogenetic tree.**

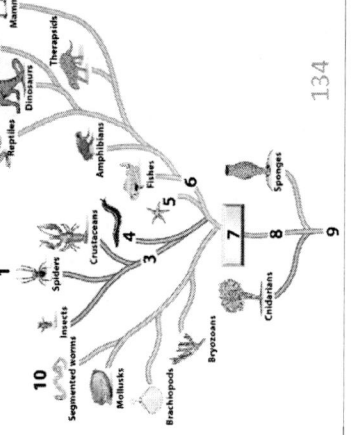

Taxonomy & Binomial nomenclature

- **Taxonomy-**It is the scientific discipline of naming and classifying organisms.
- Organisms are named using a **binomial nomenclature** where the first part is the **Genus** and second part **Species.**
- The taxonomic groups from broad to narrow are **domain, kingdom, phylum, class, order, family, genus, and species.**

Within the Domain of Eukarya

An example of the various taxonomic groups narrowing down from Kingdom to Species

Morphological and Molecular data used to create Phylogenies

- To infer phylogeny we must know the evolutionary relationships that exist between species as a result of common ancestry.

- Similarities due to common/shared ancestry are called **homologies**- and it exists at two levels

1. Morphological- Similarity in Structure
2. Molecular – Similarity in DNA sequence

Morphological Homologies

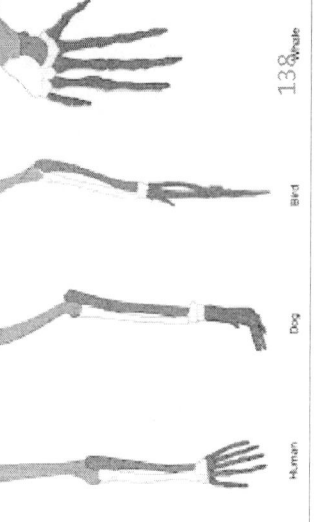

- Morpholcgical- Similarity in Structure, e.g. Arrangement of bones in the forelimbs of mammals.

- Results from shared ancestry:
- e.g. Human, Dog, Bird and Whale's forelimbs
- Also, Humans and Chimpanzee's.

Molecular Homologies

- *Molecular Homologies* – Results from similarity in genome or DNA sequence.

- This is achieved with the help of computer programs and mathematical models to analyze the DNA segments from different organisms.

- E.g.: Humans and Chimpanzees share approximately 96% of the DNA. The remaining 4% accounts for the differences that exists between us and them.

Examples of homologies at the molecular level are genes shared among organisms inherited from a common ancestor

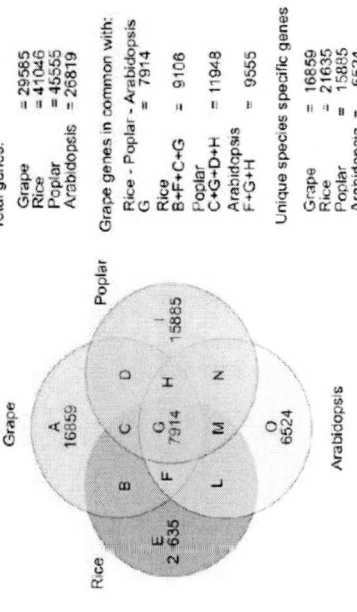

Homology vs Analogy

- In constructing a phylogeny we must not confuse Homology with Analogy and should be able to distinguish between them

- **Analogy** results from Convergent evolution, e.g. Sugar glider (marsupial) and flying squirrel.

- Analogous structures or molecular sequences that evolved independently are also called **homoplasies.**

Sugar glider and the Flying squirrel are very similar with their small rodent-like body structure

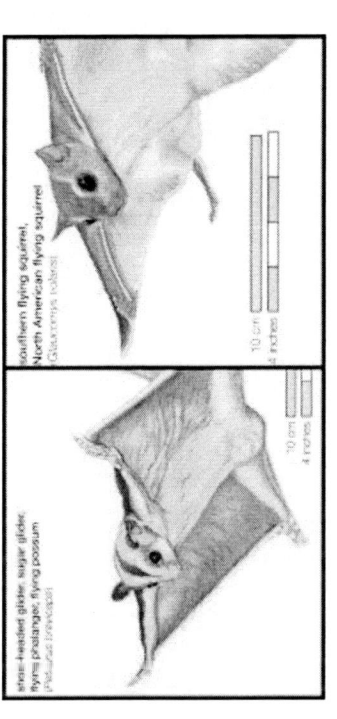

- The thin membrane that connects their forelimbs to their hind limbs is used to glide through the air in the same fashion.42

Shared characters are used to construct Phylogenetic tree

- In comparison with its ancestor, an organism has both **shared** and **unique** characteristics.

- A **shared ancestral character** is a character that originated in an ancestor, e.g. backbone in mammals is shared with other non-mammalian vertebrates.

- A **shared derived character** is an evolutionary novelty unique to a particular clade (hair in mammals is unique to mammals.)

Phylogenetic Trees for Phylogenetic bracketing

- Phylogenetic trees are used for hypotheses (educated guess) using all available data: morphological, molecular, and fossils.

- Using the data, we can make and test predictions based on the assumption that our hypothesis is correct.

- For e.g. in an approach called **Phylogenetic bracketing** we can predict that features shared by two groups of closely related organisms are present in their common ancestor and all of its descendants (clade), unless independent data indicate otherwise.

Applications of Phylogenetic bracketing

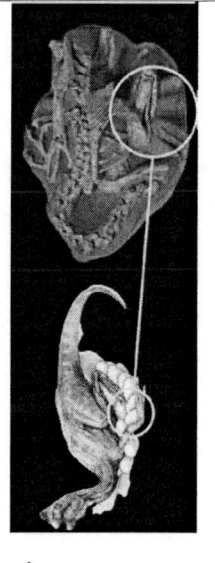

- It has been applied to infer features of dinosaurs from their descendants, e.g.: birds and crocodiles.

- Birds and crocodiles share features like brooding and four chambered hearts, which has been predicted as features of Dinosaurs.

An organism's evolutionary history in its genome

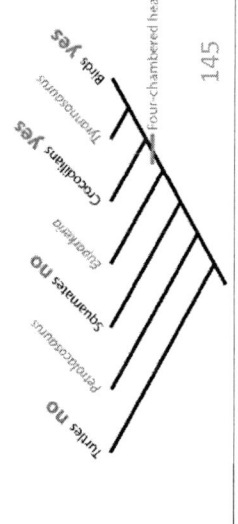

- **Orthologous genes** (b/w species) - are homologous genes that are found in different species because of speciation.

- E.g. Cytochrome C gene in Human and dogs has same function but the gene sequence has diverged.

(Paralogous genes and molecular clocks)

- **Paralogous genes** (within species) - result from gene duplication, so they are found in more than one copy in the same genome.

- Like homologous genes, duplicated genes can be traced to a common ancestor.

- E.g. Olfactory receptor gene family in humans have diverged from each other.

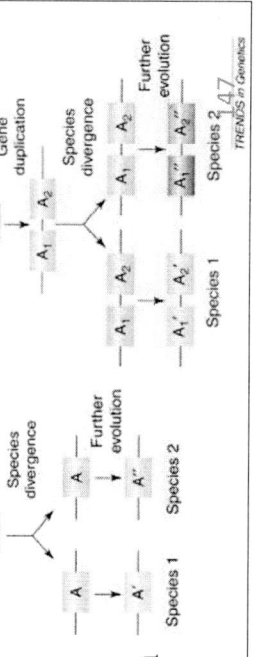

Molecular clocks help track evolutionary time

- A **molecular clock** uses constant rates of evolution in some genes to estimate the absolute time of evolutionary change

- In case of orthologous genes, nucleotide substitutions are proportional to the time since they last shared a common ancestor

- In case of paralogous genes, nucleotide substitutions are proportional to the time since the genes became duplicated

- Molecular clocks are calibrated against branches whose dates are known from the fossil record

Two Kingdoms to Three Domains

- Early taxonomists classified all species as either plants or animals.

- More recently, the three-domain system has been adopted: Bacteria, Archaea, and Eukarya.

- The three-domain system is supported by data from many sequenced genomes

Tree of Life

- The tree of life suggests that eukaryotes and archaea are more closely related to each other than to bacteria.

- The tree of life is based largely on ribosomal RNA (rRNA) genes, as these have evolved slowly.

- There have been substantial interchanges of genes between organisms in different domains

Tree of Life

Horizontal gene transfer

- **Horizontal gene transfer** is the movement of genes from one genome to another

- Using Horizontal gene transfer an organism transfers genetic information to another organism other than its offspring.

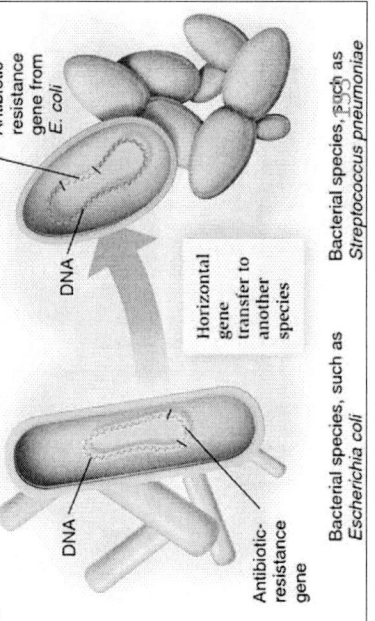

Bacterial species, such as *Escherichia coli*

Bacterial species, such as *Streptococcus pneumoniae*

Horizontal gene transfer to another species

Horizontal gene transfer

- Horizontal gene transfer also happens between organisms that are unrelated to each other. In some cases, the organisms belong to different species.

- The Horizontal gene transfer is much more common in Bacteria and Archaea but has also been studied in fungi, insects, and plants.

- E.g. Horizontal gene transfer has been seen between Millet and Rice, and also between bacteria and Yeast

154

Horizontal gene transfer can happen via

1. Transformation
2. Transduction
3. Conjugation

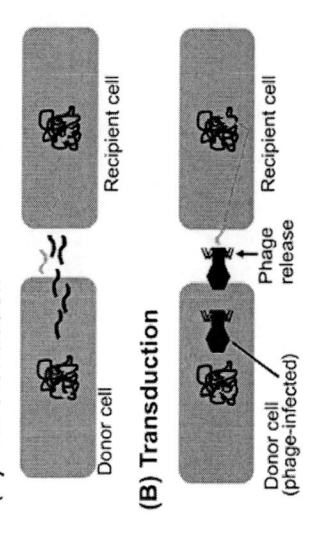

(A) Transformation

(B) Transduction

(C) Conjugation

- Comparisons of complete genomes from 3 domains show that there have been substantial movements of genes between organisms in the different domains via **Horizontal gene transfer.**

- Horizontal gene transfer is the primary reason for the spread of antibiotic resistance in bacteria

155

Horizontal Gene Transfer in Eukaryote

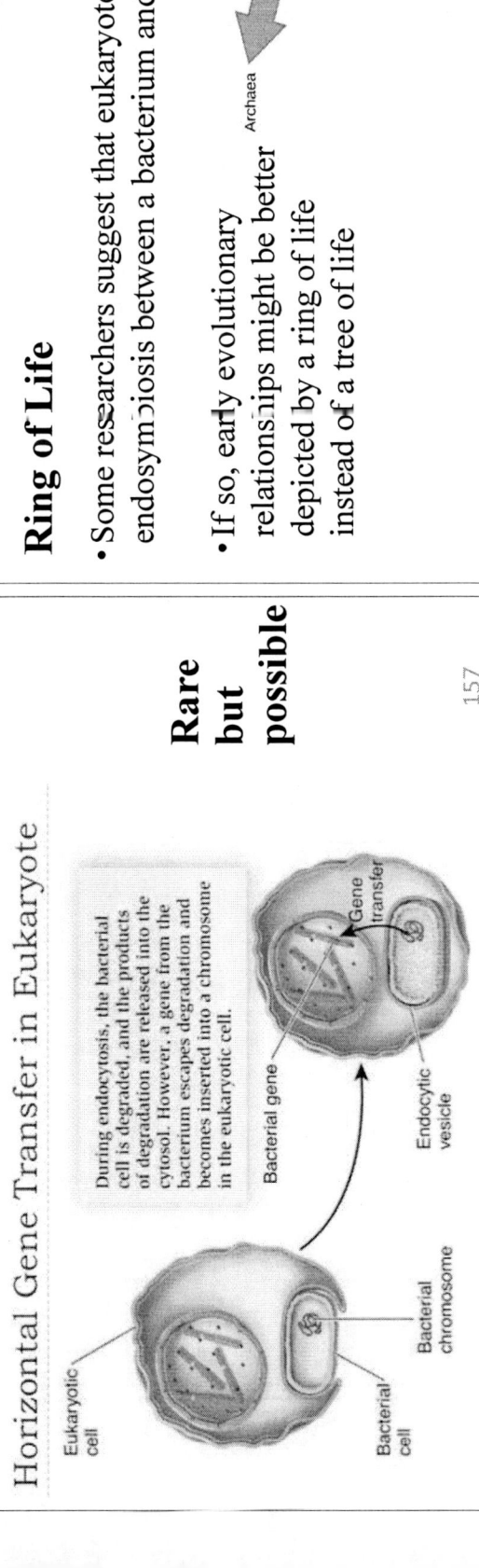

Rare but possible

Ring of Life

- Some researchers suggest that eukaryotes arose as an endosymbiosis between a bacterium and an archaea
- If so, early evolutionary relationships might be better depicted by a ring of life instead of a tree of life

6, Origin of Plants

Plants: The Lifeline of the planet

- Plants colonized land around 450 million years ago & have now diversified into roughly 400,000 species (terrestrial & marine)
- Plants supply oxygen, essential nutrients & organic macromolecules (food) to forms of life on the planed and are essential for the existence on an Ecosystem.
- Comparisons of both nuclear and chloroplast genes point to charophytes as the closest living relatives of land plants

Interesting Facts about Plants

- One kind of palm tree has a huge seed that can weigh more then 60 pounds.
- The tallest tree spices is the coast redwood, which can grow over 375 feet high.
- The leaves of the giant Amazon water lily can grow as large as 8 feet across.
- Wolffia globosa, the worlds smallest flowering plantis about the size of a grain of Rice.

Land plants evolved from green algae called Charophytes

- Four key traits that plants share with Charophytes :
1. **Rose-shaped** complexes for cellulose synthesis
2. **Peroxisome** enzymes
3. Structure of **flagellated sperm**
4. Formation of a **Phragmoplast** (complex assembly of microtubules and microfilaments before cytokinesis)

Moving to Land: Pros

- The movement onto land by Charophyte ancestors endowed them certain advantages like

1. Abundant sunlight
2. Plentiful CO_2 in the air than was dissolved in water.
3. Nutrient-rich soil
4. Very few predators(herbivores or pathogens) since the land wasn't colonized yet.

Charophytes

Moving to Land: Cons

- Despite all the advantages that land offered, it also presented challenges such as:

1. Scarcity of water
2. Lack of structural support in the air medium vs water that had more density.

Shared Derived Traits of Plants

- Four key traits appear in nearly all land plants but are absent in the Charophytes:

1. **Alternation** of generations (between Sporophtye and Gametophtye)
2. Walled **Spores** produced in **Sporangia**
3. Multicellular **Gametangia**
4. Apical **Meristems**

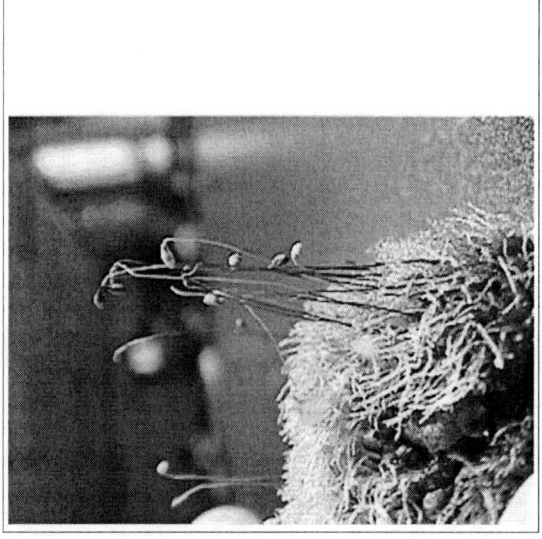

1. Alternation of Generations

- Plants alternate between two multicellular stages, **Gametophyte and Sporophyte**

- This alternation between these 2 stages makes up the plant's reproductive cycle and is called alternation of generations.

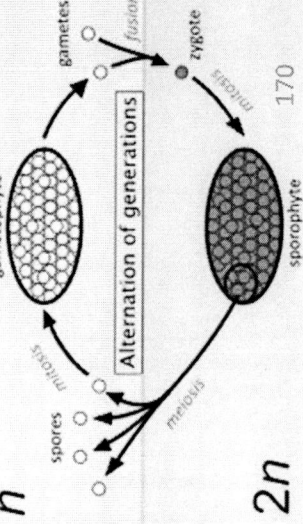

- Before we understand how the generations alternate let us first familiarize ourselves with following terms

1. **Two Reproductive cell:**   **Gametes** (Sexual reproduction)
                                        Spores (Asexual reproduction)

2. **Two multicellular stages:**   **Gametophyte** (Diploid)
                                          Sporophyte (Haploid)

3. **Two modes of cell division:** **Mitosis** (makes Diploid cell)
                                          Meiosis (makes Haploid cell)

Plant's Life Cycle

- The Fertilization of the gametes (sperm and egg) gives rise to the diploid zygote.

- Zygote then grows & divides by Mitosis and turns into a **sporophyte** (Diploid stage)

- This Sporophyte further Via Meiosis produces **Spores** (reproductive cells)

- The Spores then undergo Mitosis and make a **Gametophyte** (haploid stage)

- The gametophyte further undergoes Mitosis & makes gametes
- These gametes would then undergo fertilization and entire cycle would repeat again.

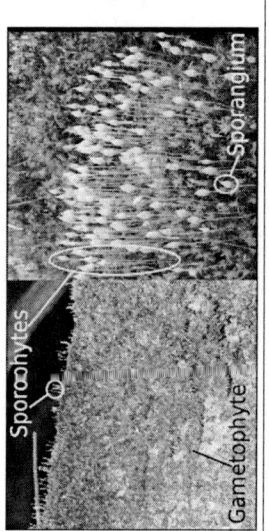

2. Spores made in Sporangia

- The sporophyte (plant stage) produces spores (reproductive cell) inside multicellular organs called **sporangia**

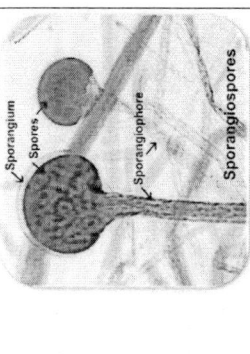
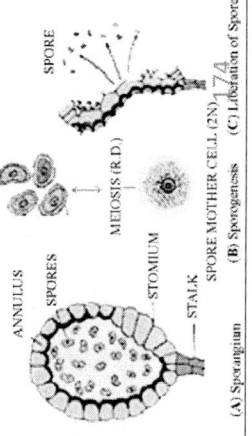
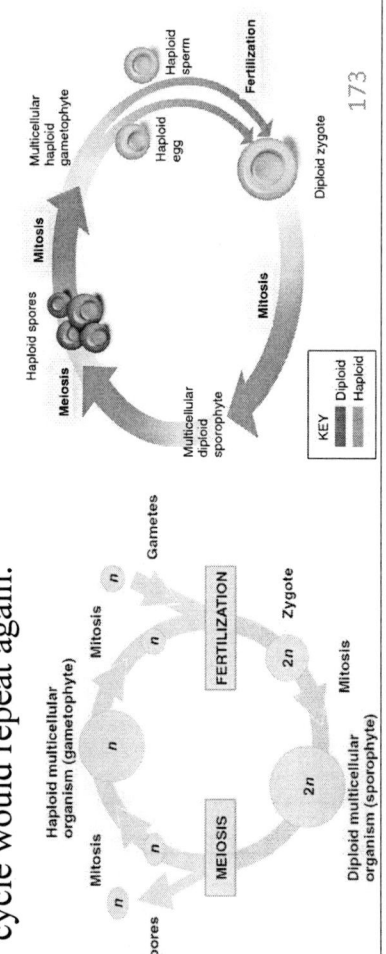

3. Gametes made in Gametangia

- The Gametangyte (plant stage) produces gametes (reproductive cell) inside multicellular organs called **Gametangia.**
- Male gametangia, **antheridia**, are the site of sperm production.
- Female gametangia, **archegonia**, produce eggs and are the site of fertilization

4. Apical Meristems

- **Apical meristems** are regions of actively dividing cells that sustain continual growth for plants.
- The cells found in Apical Meristems are undifferentiated cells also known as **Meristematic cells.**
- The cells of apical meristems differentiate into various tissues & organs of the plant and are hence analogous in function to stem cells in animals.

Two kinds of Meristems

- The 2 Meristems found in plants are shoot apical meristem & root apical meristem.

- **Shoot apical meristem** gives rise to structures like Leaves and flowers.

- While the **root apical meristem** gives rise to Roots

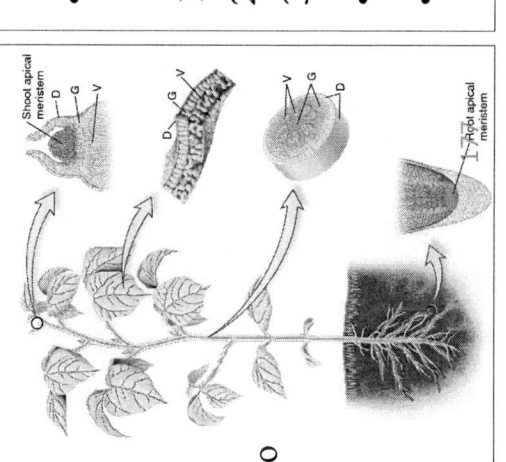

The Origin and Diversification of Plants

- Plants colonized land around 450 million years ago and diversified into 3 distinct kinds:

1. Non-vascular Plants (also called Bryophytes)
2. Seedless Vascular plants
3. Seed bearing Vascular plants

- Simplest & most primitive of these are the Nonvascular plants.
- Most complex & recent of these are the Vascular seed bearing plants

178

Non-Vascular Plants

- **Bryophytes** were the only prevalent plants for the first 100 million years of plant evolution

- Nonvascular plants have no roots, stems, or leaves, since all of these structures are made of vascular tissue (Xylem & Phloem)

- In Bryophytes, gametophytes are larger and longer-living than sporophytes. The 3 examples of Non-vascular plants

1. Liverworts   2. Hornworts   3. Mosses

179

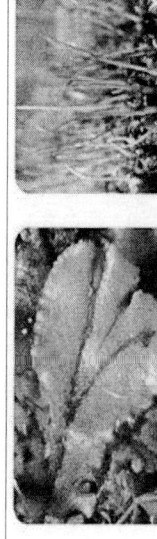

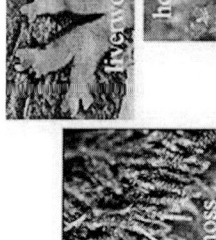

Liverwort   Hornwort   Moss

liverwort   hornwort

moss

Hornworts   Liverworts   Mosses

Vascular Plants

- Vascular plants began to diversify later and the vascular tissue allowed these plants to grow tall.
- In Vascular plants sporophytes are larger and longer-living than gametophytes
- Vascular plants have two types of vascular tissue:
  1. **Xylem** conducts most of the water and minerals
  2. **Phloem** consists of living cells and distributes sugars, amino acids, and other organic products

Examples of Vascular plants

- Examples of **Seedless** vascular plants
  1. **Lycophytes** (club mosses and their relatives)
  2. **Pterophytes** (ferns and their relatives)
- Examples of **seed bearing** vascular plants
  1. **Gymnosperms**, the "naked seed" plants, e.g. conifers
  2. **Angiosperms**, the flowering plants

Phylum Lycophyta — club mosses
Phylum Sphenophyta — horsetails
Phylum Pterophyta — fern
Phylum Psilotophyta — Whisk fern

GYMNOSPERMS

Angiosperms — Flowering Plants

The Importance of Mosses

- Mosses are capable of inhabiting diverse and extreme environments
  a) Some mosses might help retain nitrogen in the soil
  b) Mosses form organic material known as **peat** which is an important reservoir of organic carbon & is used as an energy source.

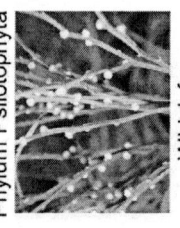

USING PEAT MOSS IN THE GARDEN

7, Origin of Seed Plants

Seeds: Nature's exceptional gift

- Seeds **transformed** the course of plant evolution, enabling the plants to become the dominant producers of the ecosystems

- This transformation of plants in their complexity further influenced animals nutrition & survival and in effect influenced the evolution of animals

- A **seed** consists of an embryo and nutrients surrounded by a protective coat

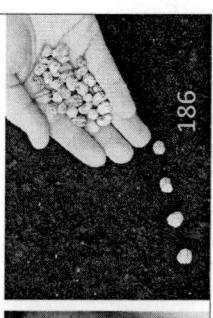
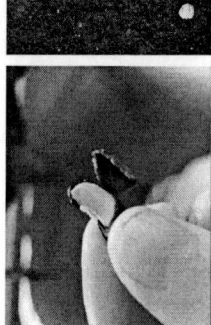

Seeds: an Adaptation for life on land

- The characteristics common to all seed bearing plants:

1. **Dominant** Sporophyte (reduced gametophyte)
2. **Heterospory** (2 kinds of spores, microspore(male) & megaspore(female)
3. **Ovules** (where the egg is made)
4. **Pollen** (where sperm is made)

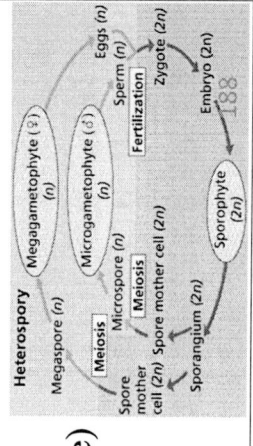

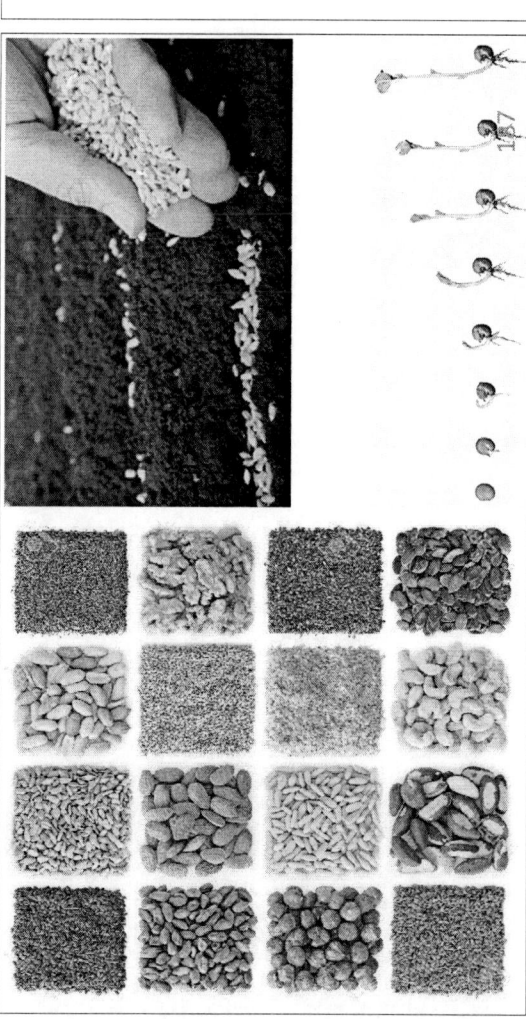

Dominant Life stage of a Seed Plant- Sporophyte

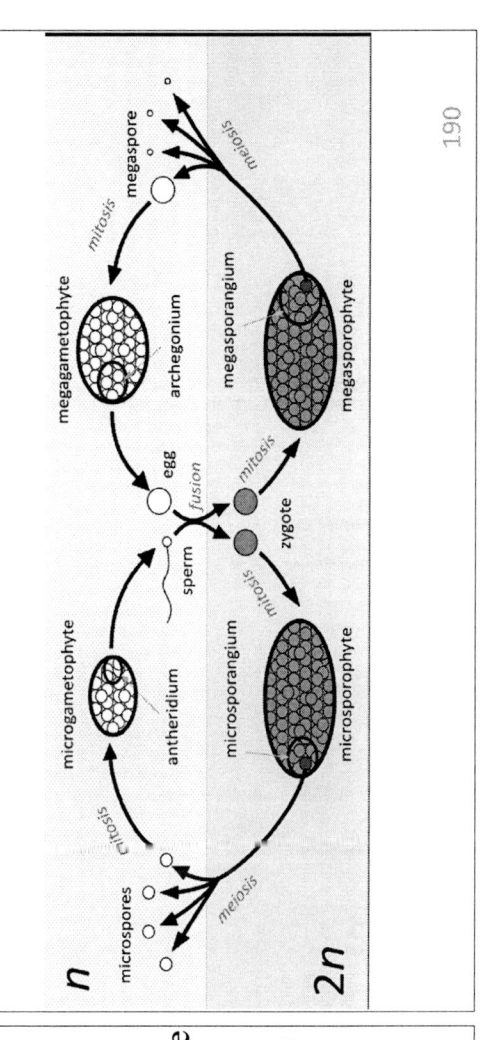

- The sporophyte (stage) produces spores (cells) in multicellular organs called **sporangia.**

- The female sporangia is referred as **megasporangium** and male as **microsporangium.**

- Megasporangium goes on to form **megaspores (female** spores) that give rise to female gametophyte (inside the ovule)

- Microsporangium on the other hand, forms **microspores** (male spore) that giver rise to male gametophyte (inside pollen grain)

Pollination

- **Pollination** is the transfer of pollen from pollen grains to the ovules, this leads to fertilization of egg within ovules.

- Pollen has the ability to travel long distances sometimes via wind, other times via insects.

- Most flowers have mechanisms to ensure **cross-pollination** between flowers from different plants of the same species.

- If a pollen grain germinates, it gives rise to a pollen tube that discharges sperm into the female gametophyte of the ovule

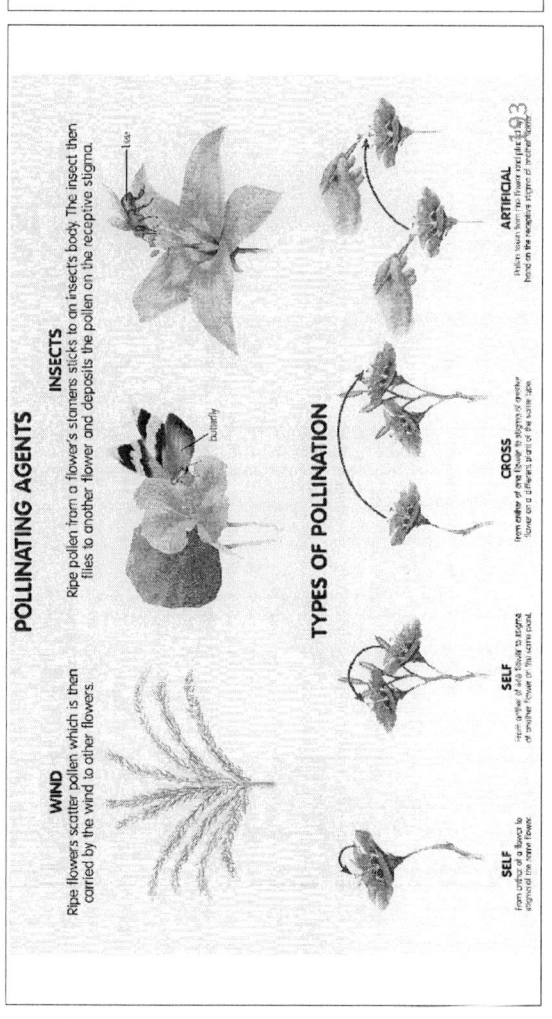

What is a seed? And How is it formed

- It is an embryo with nutrients enclosed in a protective coat.

- After successful fertilization of a sperm and egg the ovule is what become the seed. Another way to define a seed is that:

- It is a sporophyte embryo, along with its food supply, packaged in a protective coat

- The protective coat of a seed is what makes it withstand harsh conditions by keeping the embryo and nutrients safe

Seeds- The game changers

- Seed have the ability to remain dormant even for several hundred years, until conditions are favorable for germination

- Seeds are mainly dispersed by wind & animals but other modes are gravity, ballistic & water.

- Some seeds can remain in the soil seed bank for more than 50 years before germination.

- Using radioactive carbon dating the oldest documented germinating seed was found to be nearly 2000 years old

Besides seeds, additional adaptations of Angiosperms

- Some key reproductive adaptations of angiosperms include structures like flowers and fruits.

- These reproductive structure shave conferred great advantage to Angiosperms & have turned them into being the most widespread and diverse species of all.

- The species no. of Angiosperms are the highest among all of the plant kingdom at approx. 369,000 species of the total 400,000 species of plants.

Flower & Fruits

Structure of Flowers

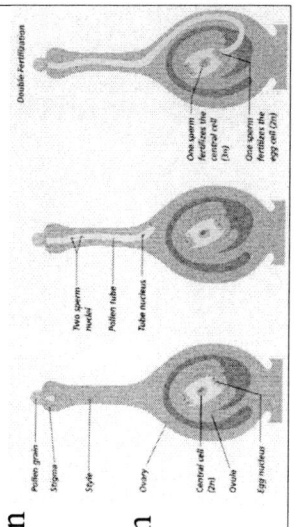

Seeds are inside the fruits.
Many seeds. | One seed.
Watermelon Kiwi | Peach Plum
Avocado | Apple Peach

Flower : a specialized shoot with four types of modified leaves

- Modified leaves of a flower:
  1. **Sepals**, enclose the flower
  2. **Petals**, are brightly colored and attract pollinators
  3. **Stamens**, produce pollen on their terminal **anthers**
  4. **Carpels**, consists of a Stigma, Style and Ovary

- Within ovary, we find **ovules**, within ovules, **embryo sac**, within embryo sac, an **egg**.

The Angiosperm Life Cycle

- Male gametophyte is contained within pollen grains.
- Female gametophyte is contained within an ovule in the ovary
- A pollen grain when lands on stigma & germinates, the result is a pollen tube, that tube then grows down within the style to reach the ovary.

Fruits: a mature ovary after fertilization

- A **fruit** is an ovary that results after fertilization of a sperm and egg in the ovule.
- After fertilization, as ovules turn into seed , the wall of the ovary thickens
- Fruits protect seeds and aid in their dispersal
- Mature fruits can be either fleshy or dry
- Seeds can be carried by wind, water, or animals to far locations

Double fertilization

- **Double fertilization** occurs when the 2 sperms discharged by pollen tube reach their targets within the ovule and unify or amalgamate with them

- 1 sperm(n) fertilizes the egg(n) and makes a Zygote (2n).

- 2nd sperm(n) combines with two nuclei (2n) in the central cell of the female gametophyte and initiates development of food-storing **endosperm**

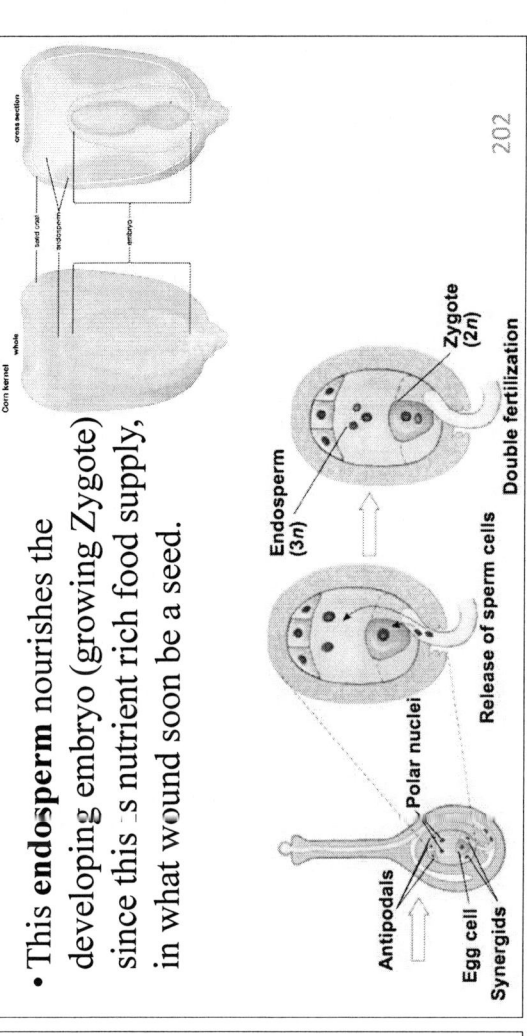

- This **endosperm** nourishes the developing embryo (growing Zygote) since this is nutrient rich food supply, in what would soon be a seed.

Human welfare & seed plants

- Plants are key sources of food, fuel and medicine

- Most of our food comes from angiosperms

- Six crops (wheat, rice, maize, potatoes, cassava, and sweet potatoes) yield 80% of the calories consumed by humans

- Many compounds of seed plants are used in medicines.

8. Plant's Growth & Differentiation

Anatomy of Plants

- Plants, like animals, have **Organ Systems**, made up of **organs**.
- These **organs** are made from different **tissues** & these tissues are a collection of specialized **cells**.
  a) **2** Organ systems: Root & Shoot system
  b) **3** Organs: Roots, Stems, & Leaves
  c) **3** Tissues: Dermal, Vascular & Ground
  d) **5** cells: Parenchyma, Collenchyma, Sclerenchyma, Tracheid's & Vessel, Sieve cells

The 2 Organ Systems

- The 2 organ systems found in plants are the **Root system** and **Shoot system**
- Shoots depends on the water and minerals absorbed by the root system
- Roots depends on sugar produced by photosynthesis in the shoot system.

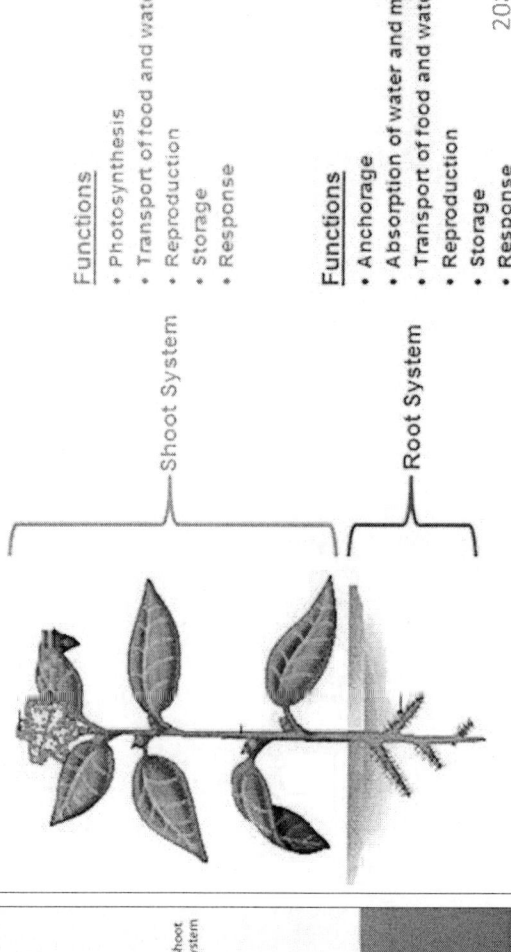

Shoot System

Functions
- Photosynthesis
- Transport of food and water
- Reproduction
- Storage
- Response

Root System

Functions
- Anchorage
- Absorption of water and minerals
- Transport of food and water
- Reproduction
- Storage
- Response

The 3 Organs

- The Three Basic Organs that make up the organs systems in a plant are:

1. **Roots:** **Absorb** water & minerals. **Anchors** plants, **Stores** organic nutrients.
2. **Stems:** **Support** to the Shoot & a **conduit** between Root and shoot
3. **Leaves:** Main site for **Photosynthesis**

Roots: organs that serve several important functions

1. **Roots absorb** water & minerals for both the root & shoot system
2. **Anchors** plants in soil & offers protection against heavy winds & floods
3. **Stores** organic nutrients as an energy reservoir.

- In most plants, absorption of water and minerals occurs near the **root hairs**, (millions of tiny root hairs increase the surface area for absorption)

Tap Roots vs Fibrous Root

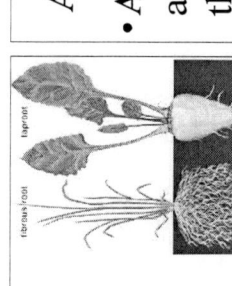

- A **taproot** system consists of one main vertical root that gives rise to **lateral roots**, or branch roots.

- A **fibrous root** is formed by thin, moderately branching **roots** growing from the stem & is a kind of **Adventitious** root.

Adventitious Roots

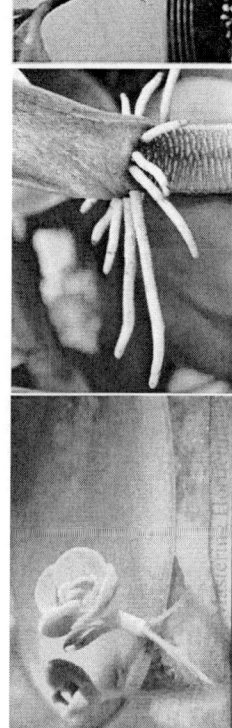

- **Adventitious roots are those that** arise from an organ other than the **root**—usually from a stem but sometimes a leaf.

Adventitious roots

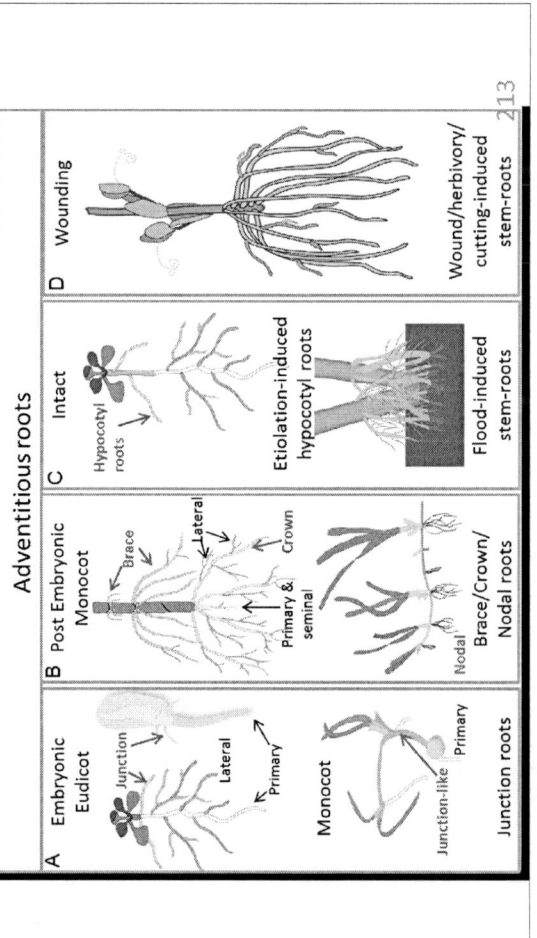

Stems: an organ consisting of nodes & internodes

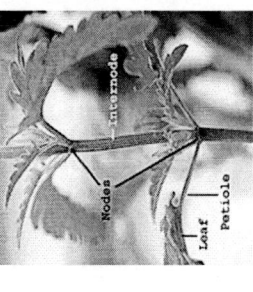

- Stem consists of an alternating system of **nodes**. (nodes are points at which leaves are attached)
- The stem segments between 2 nodes is an **Internode.**
- An **apical bud**, or terminal bud, is located near the shoot tip and causes elongation of a young shoot
- An **axillary bud** is a bud that has the potential to form a lateral shoot, or branch

---

Stems

- Stems are the main sites of growth (makes plant grow in length). It also offers support to leaves, flowers & fruits.
- It is also the main conduit between Root and Shoot system.
- The inhibition of axillary (lateral) buds by an active Apical bud is called **Apical Dominance.**
- In other words, it is a phenomenon whereby the main **central stem** of a plant is **dominant** over other side stems.

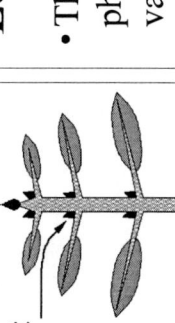

Leaves

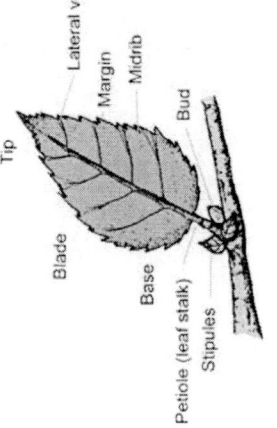

- The **leaf** is the main photosynthetic organ of most vascular plants.
- Leaves generally consist of a flattened **blade** and a stalk called the **petiole**, which joins the leaf to a node of the stem.
- Plants with 1 seed leaf(**cotyledon**) are Monocotyledons, also known as Monocots & with 2 leaves Dicotyledons or Dicots.

Classification of Plants based on Leaves

- Monocots (on left) and eudicots(on right) differ in the arrangement of **veins**, the vascular tissue of leaves.

1. **Monocots**, have one seed leaf, have parallel veins & only lateral roots.
2. **Eudicots** have two seed leaves, branching veins & a tap root along with lateral roots.

The 3 Tissues: Dermal, Vascular, and Ground

1. Dermal tissue system offers protection to plants.

- In nonwoody plants, the outer layer of cell that protect the plant from external enviorment is called **epidermis.**
- In woody plants, a protective tissues called **periderm** replace the epidermis
- Periderm is the corky outer layer of a plant stem that replaces the epidermis in older regions of stems and roots

2. Vascular tissue offers transport of water & nutrients

- Vascular system carries out long-distance transport of materials between roots and shoots.
- The two vascular tissues are xylem and phloem
- **Xylem** conveys water and dissolved minerals upward from roots into the shoots
- **Phloem** transports organic nutrients from where they are made to where they are needed

3. Ground tissue offers storage, photosynthesis & support

- Tissues that are neither dermal nor vascular are the ground tissues & are of 2 kinds based on their location, pith & cortex
- Ground tissue internal to the vascular tissue in stem is **pith.**
- Ground tissue external to the vascular tissue is **cortex.**

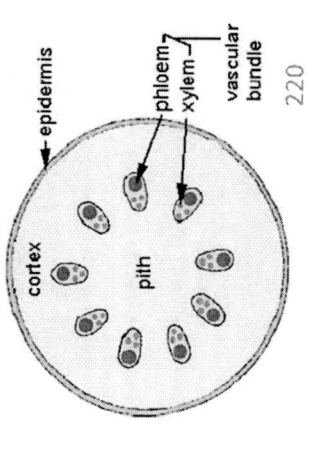

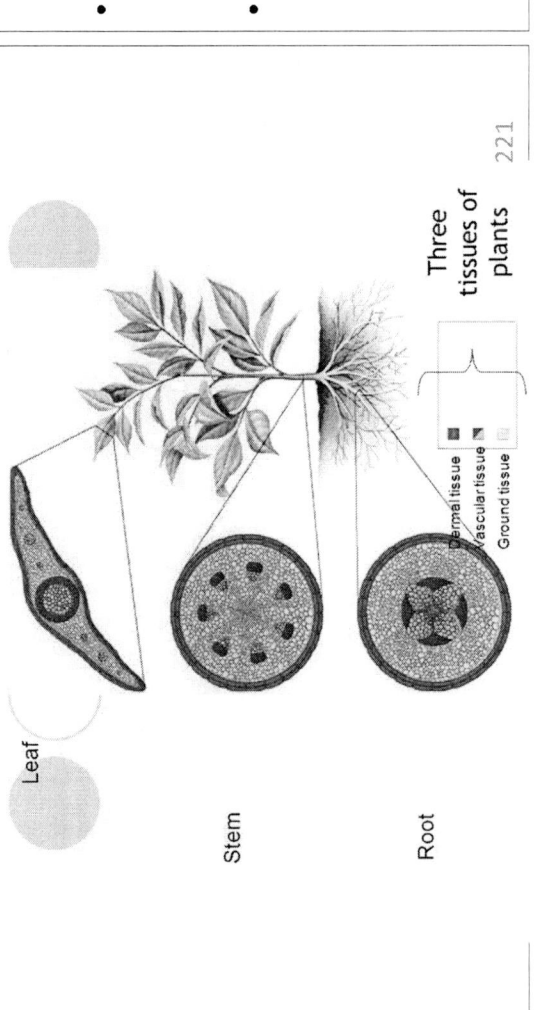

Three tissues of plants

The Five Cells

- Plants like multicellular animals, undergo cellular differentiation to create unique specialized cells that are meant to carry out a specific function

- The 5 major kinds of plant cells are

  1. Parenchyma
  2. Collenchyma
  3. Sclerenchyma
  4. Tracheid's & Vessel's
  5. Sieve elements

1. Parenchyma Cells

- Mature **parenchyma cells**

  – Have thin and flexible primary walls

  – Lack secondary walls

  – Are the least specialized

  – Perform the most metabolic functions

  – Retain the ability to divide and differentiate

Parenchyma cells with chloroplasts in it

2. Collenchyma Cells

- **Collenchyma cells** are grouped in strands and help support young parts of the plant shoot

  – They lack secondary walls

  – They have thicker and uneven cell walls

  – These cells provide flexible support without restraining growth

Collenchyma cells in stem

3. Sclerenchyma Cells

- **Sclerenchyma cells** are rigid because of thick secondary walls strengthened with lignin.
- They are dead at functional maturity.
- There are two types:
  - **Sclereids** are short and irregular in shape and have thick lignified secondary walls
  - **Fibers** are long and slender and arranged in threads

4. Tracheid's and Vessels

- **Tracheids** and **vessel elements** are the Water-conducting cells of Xylem Vascular tissue.
- It plays a vital role in the transport of water and minerals to the shoot of a plant.
- The two types of water-conducting cells, **tracheids** and **vessel elements**, are dead at maturity
- Vessel elements align end to end to form pipes called **vessels**

5. Sieve elements

- **Sieve elements** are the Sugar-conducting cells of Phloem Vascular tissue.
- It plays a vital role in the transport of Sugar to the root of the plant.
- **Sieve elements** are alive at functional maturity, though they lack organelles

Meristems generate new cells

- **Meristem** is the tissue in plants consisting of undifferentiated cells (meristematic cells) in the tips of growing stem & roots
- A Meristem gives rise to 2 kinds of cells, based on two very different roles that they are destined to fulfil
- *Initials*: Cells that remain in the meristem & keep the meristem's supply of cells in check.
- *Derivatives*: Cells that migrate away from meristem & undergo differentiation to become specialized cells of various tissues

More on Growth

- Initials are cells that keep the cell pool alive & by doing so, facilitate the migration of derivative cells away from meristem.

- The Derivatives cells are the ones that migrate away from the meristems to become specialized cell in developing tissues

- Meristems are everlasting embryonic tissue and allow for **indeterminate** growth (growth throughout life)

- Although, some plant organs cease to grow at a certain size; this is called **determinate growth**

The 2 kinds of Meristem

a) **Apical meristems** are located at the tips of roots and shoots and at the axillary buds of shoots

- Apical meristems elongate shoots & roots, a process called **Primary growth**

b) **Lateral meristems** add thickness to woody plants, a process called **secondary growth**. There are two lateral meristems-

1. The **vascular cambium**, it adds layers of vascular tissue.
2. The **cork cambium** replaces the epidermis with periderm.

Primary growth lengthens roots and shoots

- The Primary growth of plant produces the the root and shoot systems from the **apical meristems**

- Primary growth is a longitudinal growth that increase the length of the plant, unlike the secondary growth that adds thickness.

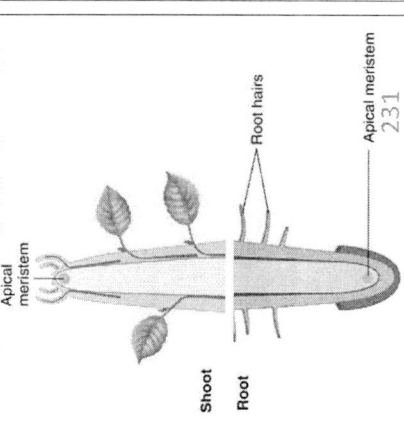

Primary Growth of Shoots

- A shoot apical meristem is a dome-shaped mass of dividing cells at the shoot tip.

- Axillary buds develop from meristematic cells at the bases of leaf primordia.

- Branch develops from axillary buds on the stem's surface

Primary Growth of Roots

A root apical meristem causes growth of roots occurs in 3 zones of cells:

1. Zone of cell division
2. Zone of elongation
3. Zone of maturation

The root tip is covered by a **root cap**, which protects the apical meristem as the root pushes through soil

Secondary Growth adds Girth to stems and roots

- Secondary growth occurs in stems and roots of woody plants but rarely in leaves.
- The **secondary plant body** consists of the tissues produced by the vascular cambium and cork cambium.
- The vascular cambium is a cylinder of meristematic cells one cell layer thick.
- The cork cambium gives rise to the secondary plant body's protective covering, or periderm

Growth, Morphogenesis, & Differentiation produce the plant body

- The **3 developmental** processes that work together to transform the fertilized egg into a plant are:

1. Growth: Cell Division & Cell Expansion
2. Morphogenesis & Pattern formation
3. Cellular Differentiation

Growth by Cell Division

a) Cell Division : increases the no. of cells, thereby causing growth.

- The plane (direction) and symmetry of cell division are immensely important in determining plant form.

- If the planes of division are parallel to the plane of the first division, a single file of cells is produced.

- If the planes of division vary randomly, asymmetrical cell division occurs

Growth by Cell Expansion

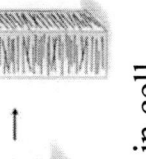

b) Cell Expansion accounts for the actual increase in plant size

- **Cellulose** Microfibrils in the cell wall restrict the direction of cell elongation

- Cytoplasmic **Microtubules** play an important role in cell division and expansion

2. Morphogenesis and Pattern Formation

- **Morphogenesis** is the development of body form and organization and is often controlled by homeotic genes.

- **Pattern formation** is the development of specific structures in specific locations

- It is determined by **positional information** in the form of signals indicating to each cell its location.

- This positional information depends on **polarity.**

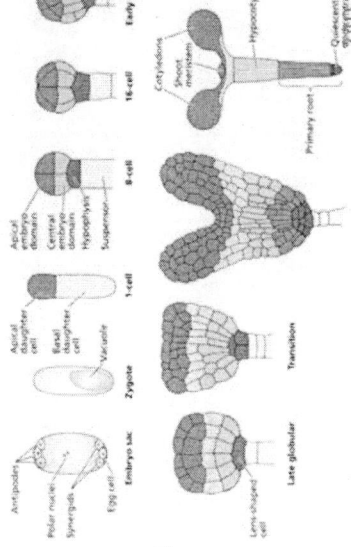

- **Polarity,** having structural or chemical differences at opposite ends of an organism, provides one type of positional information

- **Polarization is initiated** by the first asymmetrical division of the plant zygote

3. Cellular Differentiation via Gene Expression

- In cellular differentiation, cells of a developing organism synthesize different proteins and diverge in structure and function despite having a common genome.

- Cellular differentiation to a large extent depends on **positional information** and is affected by **homeotic genes.**

- Positional information underlies all the processes of development: growth, morphogenesis, and differentiation

241

Phase changes: Vegetative to Reproductive

- Plants pass through developmental phases, called **phase changes**, developing from a juvenile phase to an adult phase.

- The most obvious morphological changes typically occur in leaf size and shape

- Flower formation involves a phase change from **vegetative** growth to **reproductive** growth and is triggered by a combination of environmental cues and internal signals

242

- Transition from vegetative growth to flowering is associated with the switching on of floral Meristem identity genes

Meristem identity genes

243

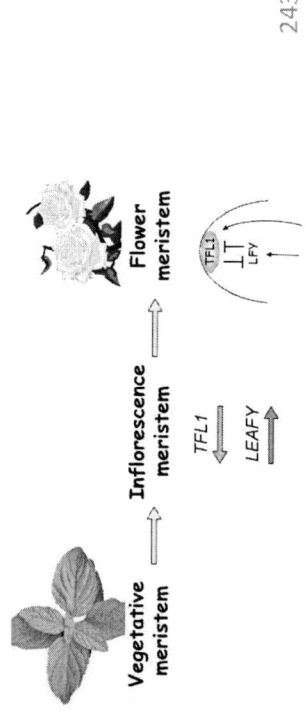

Organ identity genes

- **Organ identity genes** (plant homeotic genes) regulate the development of floral pattern

- **Mutation** in a plants organ identity gene can cause abnormal floral development

9, Acquisition & Transport of Resources

Plants acquire Resources above & below ground

- The success of plants depends on their ability to gather and conserve resources from their environment
- **Diffusion, active transport,** and **bulk flow** work together to transfer water, minerals, and sugars
- The evolution of xylem and phloem in land plants made possible the long-distance transport of water, minerals, and products of photosynthesis

Acquisition of Sunlight

- Light absorption is affected by the **leaf area index**, the ratio of total upper leaf surface of a plant divided by the surface area of land on which it grows
- **Phyllotaxy**, the arrangement of leaves on a stem, is specific to each species

Acquisition of Water & Minerals

- Soil is a resource that's rich in nutrients & with help of certain microorganisms, plants make the most of it.
- Roots and the soil fungi form symbiotic associations called **mycorrhizae**

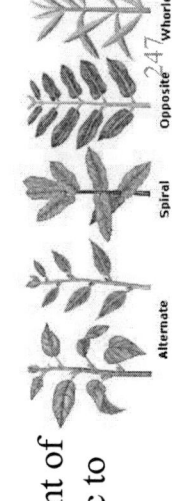

- Mutualisms with fungi played a very significant role in helping plants colonize land

Short-distance Diffusion & Active transport

- Transport begins with the absorption of resources by plant cells
- The movement of substances into and out of cells is regulated by selective permeability.
- **Diffusion** across a membrane is **passive**, while the pumping of solutes across a membrane is active and requires energy.
- Most solutes pass through **transport proteins** embedded in the cell membrane

- The most important transport protein for active transport is the **proton pump**
- Proton pumps in plant cells create a hydrogen ion gradient that is a form of potential energy that can be harnessed to do work
- This hydrogen ion gradient contributes to a voltage known as **membrane potential**, that's capable of doing work

- Plant cells use **energy** stored in **the proton gradient** and membrane potential **to transport** many different solutes
- In the mechanism of **Cotransport**, a transport protein couples the diffusion of one solute to the active transport of another

Diffusion of Water (Osmosis)

- To survive, plants must balance water uptake and loss
- **Aquaporins** are transport proteins in the cell membrane that allow the passage of water.
- **Osmosis** determines the net uptake or water loss by a cell and is affected by solute concentration and pressure
- **Water potential** is a measurement that combines the effects of solute concentration and pressure

Water potential - Ψ

- Water potential determines the direction of movement of water
- Water flows from regions of higher water potential to regions of lower water potential
- Water potential is abbreviated as Ψ and measured in units of pressure called **megapascals (MPa)**
- $\Psi = 0$ MPa for pure water at sea level and room temperature

---

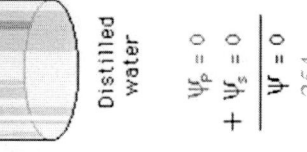

Distilled water

$\Psi_P = 0$
$+ \Psi_S = 0$
$\overline{\Psi = 0}$

- Both solute concentration and pressure affect water potential.
- The **solute potential** (Ψ_S) of a solution is proportional to the number of dissolved molecules
- Solute potential is also called **osmotic potential**
- **Pressure potential** (Ψ_P) is the physical pressure on solution

---

Water potential = (Ψ_S) + (Ψ_P)

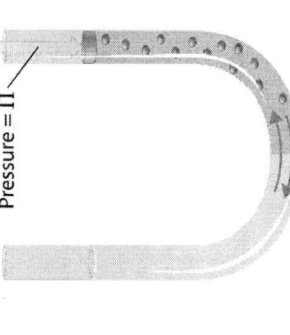

(a) Initial state — Pure water, Solution, Glucose, Semipermeable membrane

(b) Equilibrium — Osmotic pressure, Π

(c) External pressure applied — Pressure = Π

Water potential affects the uptake & loss of water

- If a flaccid cell is placed in a solution with a lower solute concentration, the cell will gain water and become **turgid.**
- If the same **flaccid** cell is placed in a higher solute concentration, the cell will lose water and undergo **plasmolysis**
- **Turgor pressure** is the pressure exerted by the plasma membrane against the cell wall.
- Turgor loss as in plasmolysis causes **wilting of plants,** which can be reversed later if the plant is watered

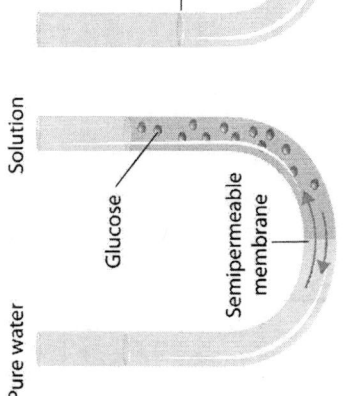

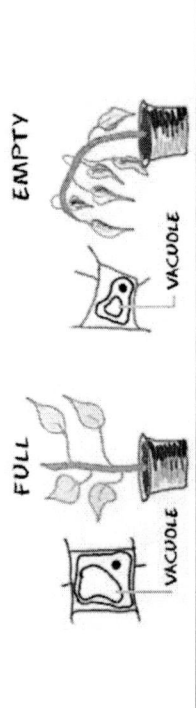

Compartmental structures of plant cells that regulate water

1. The **plasma membrane** directly controls the traffic of molecules into and out of the cell

2. Another major structure in most mature plant cells is the central **vacuole**, a large organelle that stores water & nutrients.

Three Major Pathways of Transport

• Water & minerals can travel through a plant by **3** routes:

1. **Symplastic** route: via the continuum of cytosol

2. **Apoplastic** route: via the cell walls & extracellular spaces

3. **Transmembrane** route: out of one cell, across a cell wall, & into another cell

Symplastic, Apoplastic & Transmembrane route

Apoplast is the continuum of cell walls & extracellular spaces

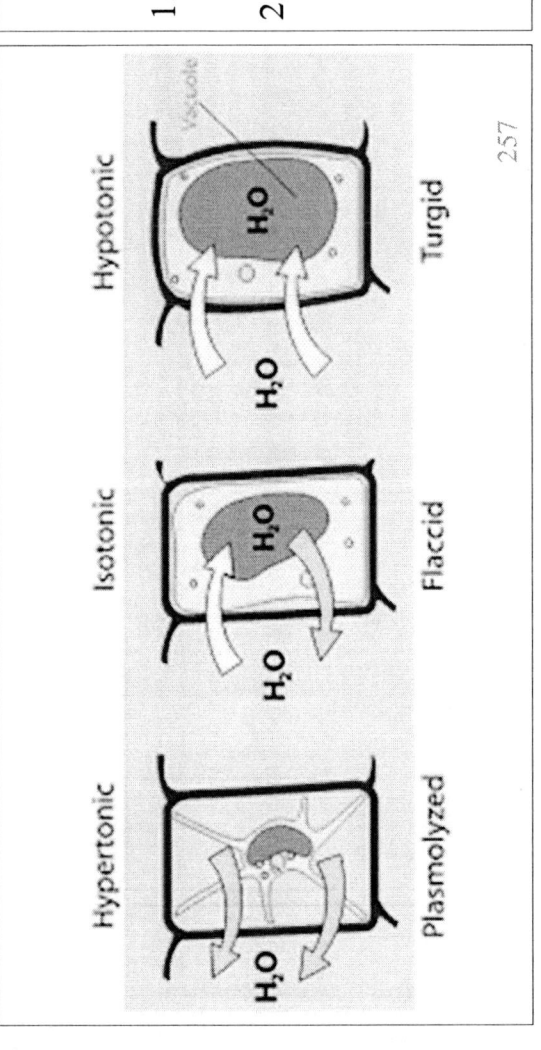

Symplast is the continuum of cytoplasm between cells & is connected by channels called **plasmodesmata**

Long-distance transport by bulk flow

- Efficient long distance transport of fluid requires **bulk flow**, the movement of a fluid driven by pressure.

- Water and solutes move together through tracheids and vessel elements of **xylem**, and sieve-tube elements of **phloem**

- Efficient movement is possible because mature tracheids and vessel elements have no cytoplasm, and sieve-tube elements have few organelles in their cytoplasm

Water & Minerals are transported from Roots to Shoots

- Plants can move a large volume of water from their roots to shoots.

- Most water and mineral absorption occurs near root tips, where the epidermis is permeable to water and root hairs are located.

- Root hairs account for much of the surface area of roots.

- After soil solution enters the roots, the extensive surface area of cortical cell membranes enhances uptake of water & minerals

---

- The **endodermis** is the innermost layer of cells in the root cortex

- It surrounds the vascular cylinder and is the **last checkpoint** for selective passage of minerals from the cortex into the vascular tissue

- Water can **cross the cortex** via the Symplast or Apoplast.

- The waxy **Casparian strip** of the endodermal wall blocks apoplastic transfer of minerals from the cortex to the vascular cylinder

Casparian Strip

Bulk Flow driven by Negative Pressure in the Xylem

- Plants lose a large volume of water from **transpiration**, the evaporation of water from a plant's surface.

- Water is replaced by the bulk flow of water and minerals, called **xylem sap,** from the roots to the stems and leaves.

- Water is pulled upward by a **negative pressure** called **transpiration pull** in the xylem

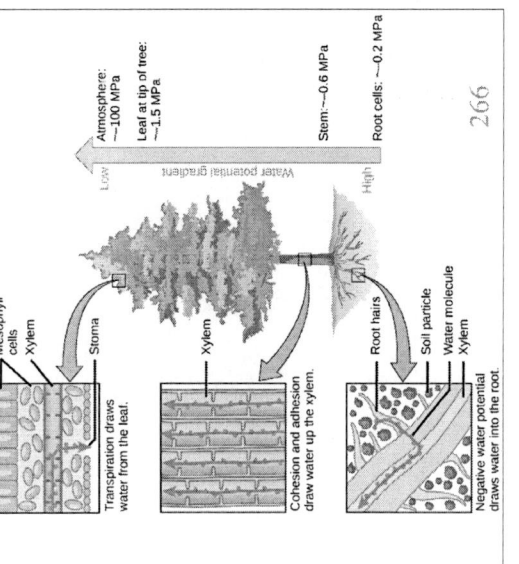

Negative Pressure in Xylem

- **Transpiration** produces negative pressure (tension) in the leaf, which exerts a pulling force on water in the xylem, pulling water into the leaf.

- The transpirational pull on xylem sap is transmitted all the way from the leaves to the root tips.

- Transpiration lowers water potential in leaves, and this generates negative pressure (tension) that pulls water up through the xylem

- There is no energy cost to bulk flow of xylem sap

The role of Cohesion & Adhesion in the ascent of Xylem Sap

- Transpirational pull is facilitated by **cohesion** of water molecules to each other & their **adhesion** to cell walls.

- The movement of xylem sap against gravity is thus maintained by the **transpiration-cohesion-tension** mechanism

Movement of water by

Transpiration

Cohesion & Adhesion

Osmosis

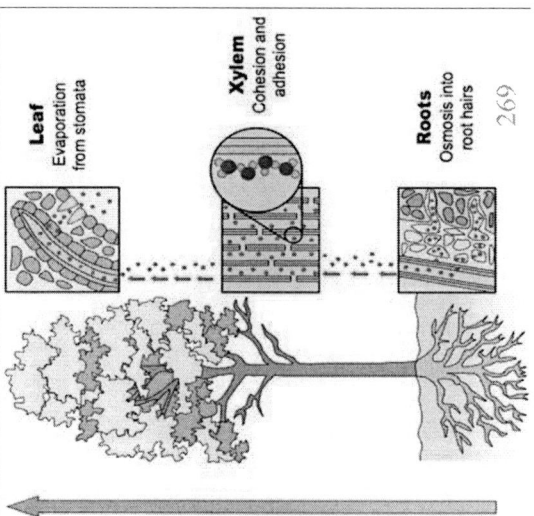

Bulk Flow driven by Positive Pressure in Phloem

- Sap moves through a sieve tube of the Phloem by bulk flow driven by positive pressure.
- In this positive pressure, the movement is in a downward direction from source(leaves & stem) to sink(roots).
- The positive pressure is facilitated by the gravitational pull that acts upon the phloem sap(sugar solution) & helps in diffusing it from its high to its low concentration.(leaf to root)

Positive Pressure in Phloem

&

Negative Pressure in Xylem

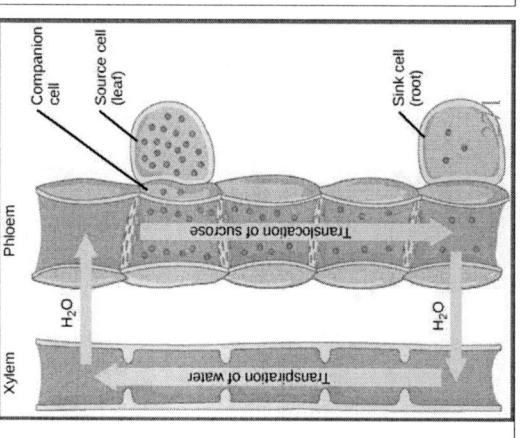

Stomata help Regulate the rate of transpiration

- Leaves generally have broad surface areas and high surface-to-volume ratios.
- These characteristics increase photosynthesis and increase water loss through stomata.
- About 95% of the water a plant loses escapes through stomata (also called stoma).
- Each stoma is flanked by a pair of **guard cells**, which control the diameter of the stoma by changing shape

Closing & Opening of stomata by deflating inflating the Guard cell

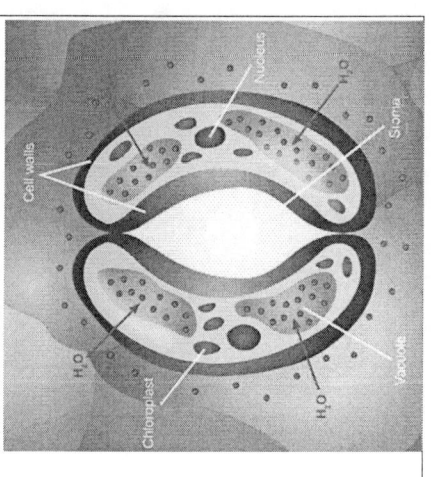

- Stomatal opening at dawn is triggered by light, CO_2 depletion, and an internal "clock" in guard cells.

- Changes in turgor pressure open and close stomata.

- These result primarily from the reversible uptake and loss of **potassium ions** by the **guard cells**

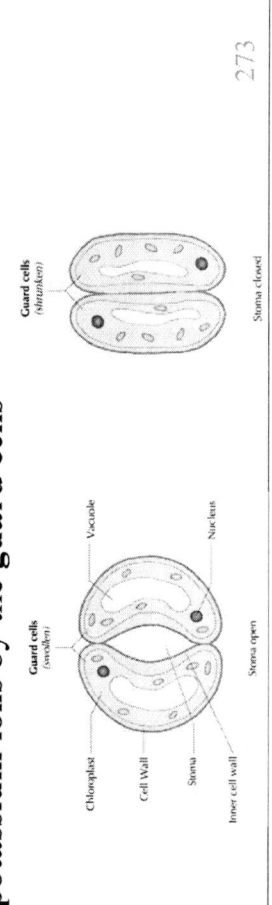

Effects of Transpiration on Wilting

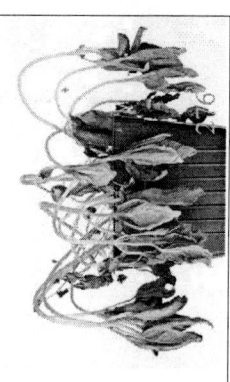

- Generally, stomata open during the day and close at night to minimize water loss.

- Plants lose a large amount of water by transpiration and if the lost water is not replaced the plant will **wilt**

How do stomata open during the day?

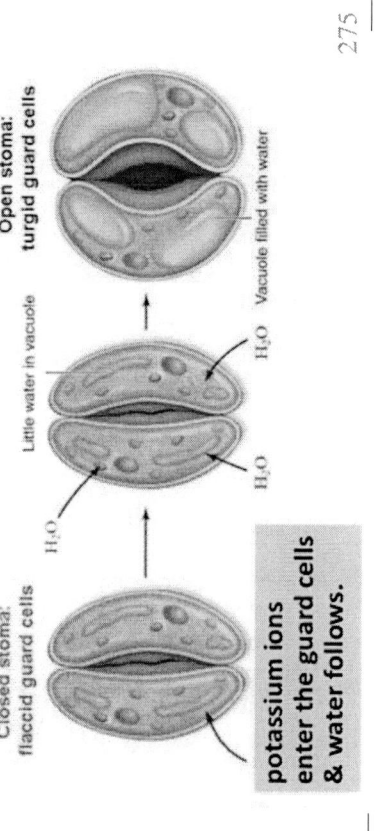

potassium ions enter the guard cells & water follows.

Effects of Transpiration Leaf Temperature

- Transpiration also results in evaporative cooling which can lower the temperature of a leaf.

- This **prevents denaturation of** various enzymes involved in photosynthesis and other metabolic processes

Adaptations that reduce Water Loss

- **Xerophytes** are plants adapted to hot, arid climates.

- They have leaf modifications that reduce the rate of transpiration.

- Some plants use a specialized form of photosynthesis called Crassulacean acid metabolism (**CAM**) where stomatal gas exchange occurs at **night**

Sugars transported from leaves to sites of use & storage

- The products of photosynthesis are transported through phloem by the process of **translocation.**

- **Phloem sap** is an aqueous solution that is high in sucrose & it travels from a sugar source to a sugar sink.

- A **sugar source** is an organ that is a net producer of sugar, such as mature leaves

- A **sugar sink** is an organ that is a uses or stores sugar, such as a tuber

Translocation of sucrose   Sugar sink: Tuber

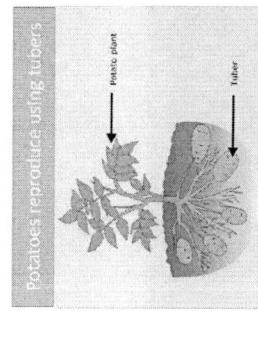 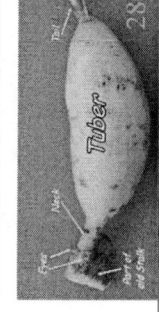

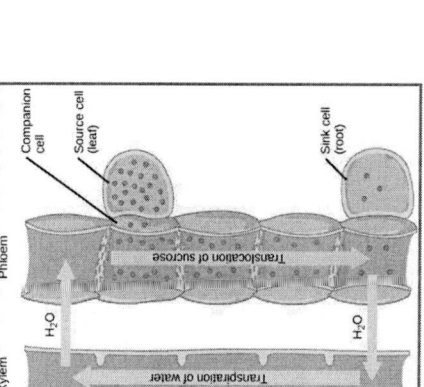

10. Angiosperms

Reproduction cycle of Angiosperms

- Diploid (2n) sporophytes produce spores (n) by meiosis; these spores grow into haploid (n) gametophytes.
- Gametophytes(n) produce gametes (n) by mitosis; later by **fertilization** gametes make Zygote.
- The Zygote undergoes Mitosis to make a sporophyte
- In angiosperms, the sporophyte is the dominant generation.
- The gametophytes are reduced in size and depend on the sporophyte for nutrients

The 3 unique features of the angiosperm life cycle are Flowers, double Fertilization, & Fruits

- The angiosperm life cycle is characterized by "three Fs": flowers, double fertilization and fruits
- Flowers are the reproductive shoots of angiosperm sporophyte.
- **4 floral organs of a Flower: Sepal, Petal, Stamen & carpel.**
- **Complete flowers** contain all four floral organs.
- **Incomplete flowers** lack one or more floral organs, for example stamens or carpels

Angiosperm Flower

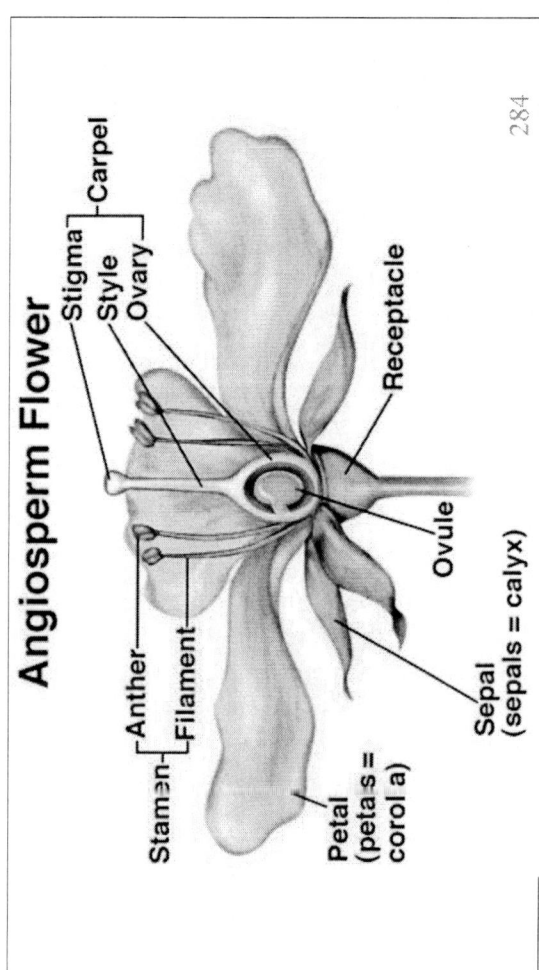

Flower Structure and Function

- A stamen consists of a filament topped by an **anther** with pollen sacs that produce pollen.

- A carpel has a long **style** with a **stigma** on which pollen may land and at the base of the style is an **ovary** containing one or more **ovules.**

- A single carpel or group of fused carpels is called a **pistil.**

Sperm made from Pollen Grains & Egg in Ovule

- Pollen develops from **microspores** within the microsporangia, or pollen sacs, of anthers

- If pollination succeeds, a **pollen grain** produces a **pollen tube** that grows down into the ovary and discharges sperm.

- The pollen grain consists of the two-celled male gametophyte and the spore wall

- Within an ovule, **megaspores** are produced by meiosis and develop into **embryo sacs**, the female gametophytes

Pollination

- In angiosperms, **pollination** is the transfer of pollen from an anther to a stigma.

- Pollination can happen by wind, water, insects, bird or humans.

- After landing on a receptive stigma, a pollen grain produces a pollen tube that extends in the style & heads towards the ovary

- The pollen tube then discharges **2 sperms** that reach the embryo sac within the ovule of the ovary

Double Fertilization: Makes Egg & Endosperm

- **Double fertilization** results from the discharge of two sperm from the pollen tube into the embryo sac.

- One sperm fertilizes the egg & other combines with two polar nuclei's, giving rise to the triploid (3n) food-storing **endosperm**

- After double fertilization, each ovule(contains egg) develops into a seed & the ovary develops into a fruit.

- Endosperm stores nutrients that can be used by the seedling

Double Fertilization

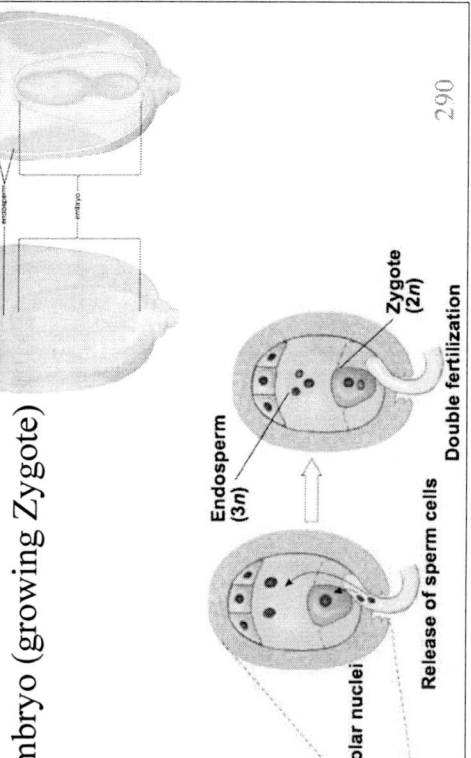

Double Fertilization

- The **endosperm** nourishes the developing embryo (growing Zygote)

Seed Dormancy & it's advantage

- The embryo and its food supply are enclosed by a hard, protective **seed coat.**

- The seed enters a state of **dormancy** which is an Adaptation for tough times

- Seed dormancy increases the chances that germination will occur at a time and place most advantageous to the seedling

- The breaking of seed dormancy often requires environmental cues, such as temperature or lighting changes

Seed Germination and Seedling Development

- Germination depends on **Imbibition**, the uptake of water due to low water potential of the dry seed

- The radicle (embryonic root) is the first to emerge. Next, the shoot tip breaks through the soil surface.

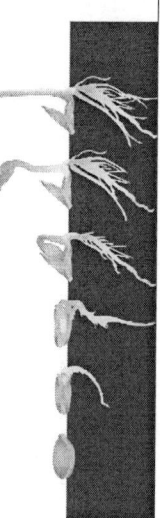

Fruit Form and Function

- A fertilized ovary become a Fruit.
- Fruit protects the enclosed seeds and aids in seed dispersal by wind or animals
- A fruit may be classified as dry, if the ovary dries out at maturity, or fleshy, if the ovary becomes thick, soft, and sweet at maturity
- Fruit dispersal mechanisms can happen via water, wind and animals

a) In **Monocots** like maize and other grasses, the coleoptile pushes up through the soil (diagram on left)

b) In **Eudicots**, a hook forms & growth pushes the hook above ground. The hook straightens & pulls cotyledons & shoot up

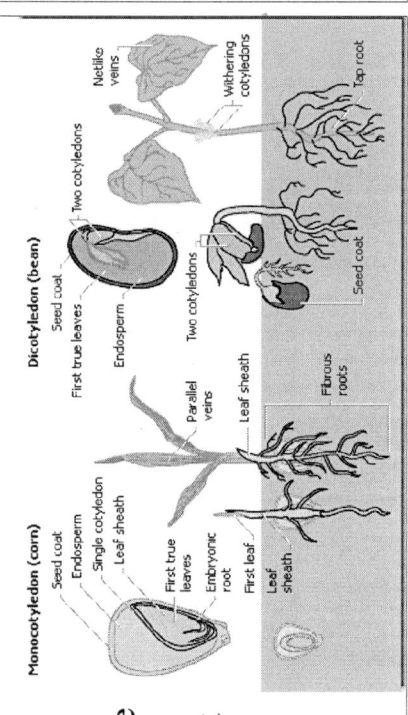

Sexual & Asexual reproduction of Angiosperms

- Many angiosperm species reproduce both asexually & sexually
- **Sexual** reproduction results in offspring that are genetically different from their parents
- **Asexual** reproduction results in a clone of genetically identical organisms. One way of asexual reproduction is **Fragmentation**.

Fragmentation, separation of a parent plant into parts that develop into whole plants.

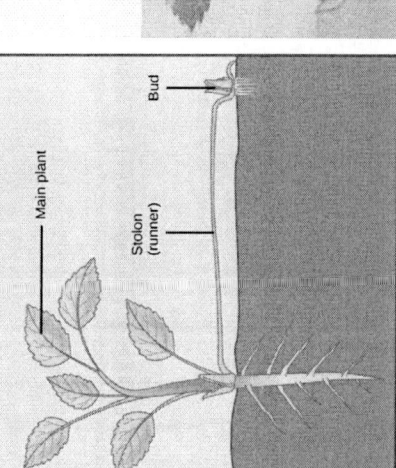

Pros & Cons of Asexual Vs Sexual Reproduction

- Asexual reproduction is also called **vegetative reproduction.**

- Asexual reproduction can be beneficial to a successful plant in a stable environment

- However, a clone of plants is vulnerable to local extinction if there is an environmental change

- Sexual reproduction generates genetic variation that makes evolutionary adaptation possible. However, only a fraction of seedlings survive

Mechanisms that Prevent Self-Fertilization

- Many angiosperms have mechanisms that make it difficult or impossible for a flower to self-fertilize.

1. The most common is **self-incompatibility**, a plant's ability to reject its own pollen by blocking growth of pollen tube.

2. Some species have stamen & carpel on separate plants

3. Others have stamens and carpels that mature at different times or are arranged such that it prevents self fertilization

Vegetative Propagation

- Humans have devised methods for **asexual propagation** of angiosperms.

1. Many kinds of plants are asexually reproduced from plant fragments called **cuttings.**

2. Another kind of vegetative propagation results from the use of **Test-tube cloning** & related techniques, 2 examples:
      a) Transgenic plants   **b) Protoplast fusion**

Cuttings

- A plant **cutting** is a piece of a plant (usually stem or root) that is used in vegetative (asexual) propagation.

- A stem cutting produces new roots, and a root cutting produces new stems.

- Some plants can be grown from leaf pieces, called leaf cuttings, which produce both stems and roots.

- At the cut end of the shoot a mass of dividing, undifferentiated cells called **Callus** forms (site where adventitious roots develop)

2. Test-Tube Cloning & Related Techniques

a) Transgenic plants are genetically modified (GM) to express a gene from another organism

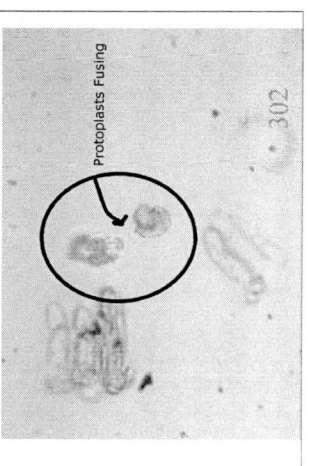

2. Test-Tube Cloning & Related Techniques

b) Protoplast fusion is used to create hybrid plants by fusing protoplasts. (which are plant cells with their cell walls removed)

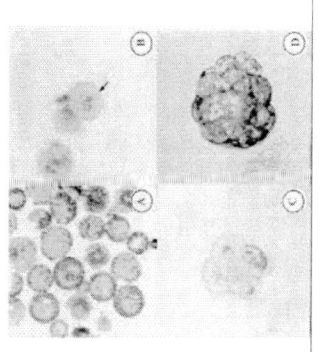

Humans modify crops by Breeding and Genetic engineering

a) Hybridization is common in nature and has been used by farmers to introduce new genes to plants (by artificial selection). E.g. : Maize

- Furthermore. plants with beneficial **mutations** are used in breeding experiments.

b) Use of GMO (Genetically modified organisms) in agriculture and industry. Genetically modified organisms include bacteria, yeast, plants, fish, and mammals.

Reducing World Hunger and Malnutrition

- Genetically modified plants may increase the quality and quantity of food worldwide
- Transgenic crops have been developed that:
  – Produce proteins to defend them against insect pests
  – Tolerate herbicides
  – Resist specific diseases

Growing Concerns: GMO Crops Worldwide

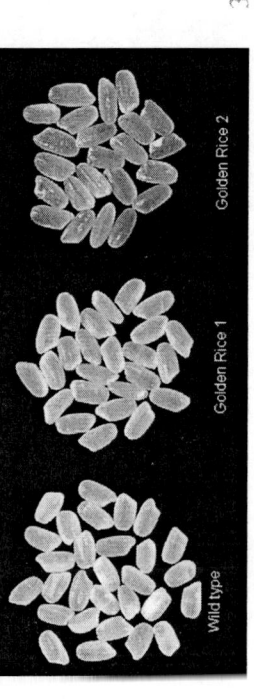

- Using Transgenic crops, the nutritional quality of plants is being improved
- "Golden Rice" is a transgenic variety being developed to address vitamin A deficiencies among the world's poor

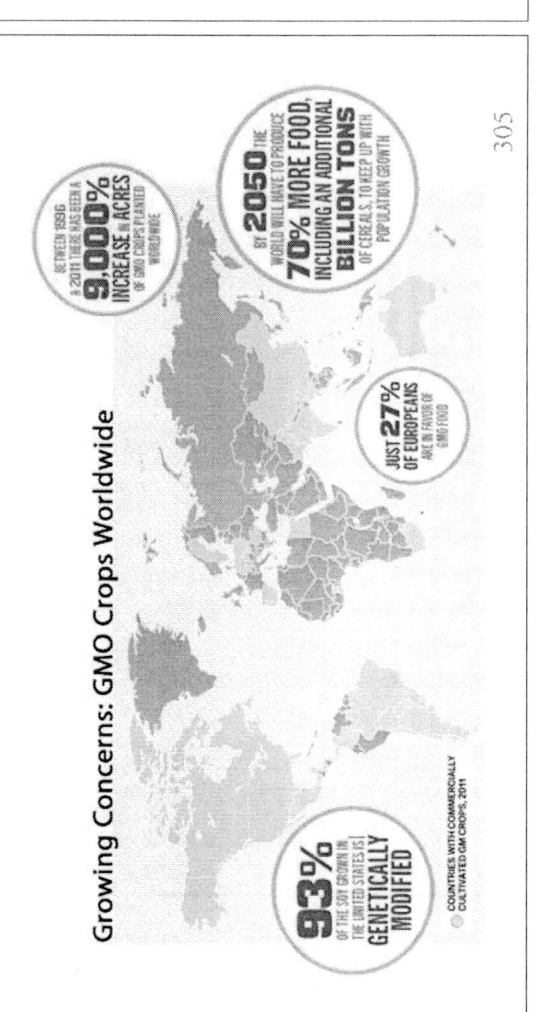

Wild type
Golden Rice 1
Golden Rice 2

Biofuels

Reducing Fossil Fuel Dependency

- **Biofuels** are made by the fermentation and distillation of plant materials such as cellulose
- Biofuels can be produced by rapidly growing **crops**
- Biofuels from plant biomass would reduce the net emission of greenhouse gas CO_2
- However, critics of biofuel technology estimate that it may require more energy to produce biofuels than would be produced from combustion of these products.

The Debate over Transgenic crops

- **Risks** associated with **Transgenic** crops (GMO) are as follows:

1. *Issues of Human Health* - Genetic engineering may transfer allergens from a gene source to a plant used for food

2. *Effects on Nontarget Organisms* - GM crops might have unforeseen effects on nontarget organisms

3. *Transgene Escape* - Introduced genes escaping into weeds through crop-to-weed hybridization

Risks associated with Transgenic crops (GMO)

11. Plant Responses

Cell signaling: Reception, Transduction & Response

1. **Reception:** Internal and external signals are detected by **receptors**, proteins that change in response to specific stimuli.

2. **Signal transduction pathway** leads to regulation of one or more cellular activities.

3. **Response** in most cases, increases the activity of the enzymes either by transcriptional regulation or by post-translational modification

Cell Signaling

Cell's Response- Transcriptional Regulation

- Specific transcription factors bind directly to specific regions of DNA and control transcription of genes.

- Positive transcription factors are proteins that *increase* the transcription of specific genes.

- Negative transcription factors are proteins that *decrease* the transcription of specific genes

Cell's Response - Post-Translational Modification of Proteins

- Post-translational modification involves modification of existing proteins in the signal response.

- Modification often involves the phosphorylation of specific amino acids

Plant hormones coordinate growth & development

- **Hormones** are chemical signals that coordinate different parts of an organism

- In general, hormones control plant growth and development by affecting the division, elongation, and differentiation of cells

- Plant hormones are produced in very low concentration, but a minute amount can greatly affect growth and development of a plant organ. The 5 main hormones of plants are:

1. Auxin and its effects

- The term **auxin** refers to any chemical that promotes **elongation** of coleoptiles

1. Auxin is involved in **root formation and branching**

2. Auxin affects secondary growth by **inducing cell division** in the vascular cambium and influencing differentiation of secondary xylem

2. Cytokinins and its effects

1. Control of Cytokinesis (cell Division) and Differentiation

- Cytokinins are produced in actively growing tissues such as roots, embryos, and fruits
- Cytokinins work together with auxin to control cell division and differentiation

2. Anti-Aging Effects

- Cytokinins retard the aging of some plant organs by inhibiting protein breakdown & stimulating RNA and protein synthesis

3. Gibberellins and its effects

1. **Stem Elongation-** Gibberellins stimulate growth of leaves and stems. In stems, they stimulate cell elongation & division.

2. **Fruit Growth-** In many plants, both auxin and gibberellins must be present for fruit to set.

3. **Germination-** After water is imbibed, release of gibberellins from the embryo signals seeds to germinate

4. Abscisic acid (ABA) and its effects

1. **Seed dormancy-** Seed dormancy ensures that the seed will germinate only in optimal conditions.

- In some seeds, dormancy is broken when ABA is removed by heavy rain, light, or prolonged cold.

2. **Drought tolerance-** ABA is the primary internal signal that enables plants to withstand drought

5. Ethylene and its effects

1. **Senescence-** induces the programmed cell death (**apoptosis**)

2. Leaf **Abscission-** Falling of leaves during autumn

3. Fruit **Ripening:** Burst of ethylene in a fruit triggers the ripening

4. Response to **mechanical stress:** Ethylene induces the **triple response** to combat mechanical stress, this allows a growing shoot to avoid obstacles

The **triple response** consists of a slowing of stem elongation, a thickening of the stem, and horizontal growth

1. Seedling stem stops elongating.
2. Seedling stem thickens.
3. Seedling stem bends, forming a hook.
Hook

Ethylene concentration (parts per million)
0.00 0.10 0.20 0.40 0.80

Phototropism

- Any response resulting in curvature of organs toward or away from a stimulus is called a **tropism** and when the stimulus is light its **phototropism.**

- Tropisms are often caused by hormones.

- Experiments on **phototropism**, found that a grass seedling could bend towards light only if the tip of the coleoptile(tip) was present.

Phototropism

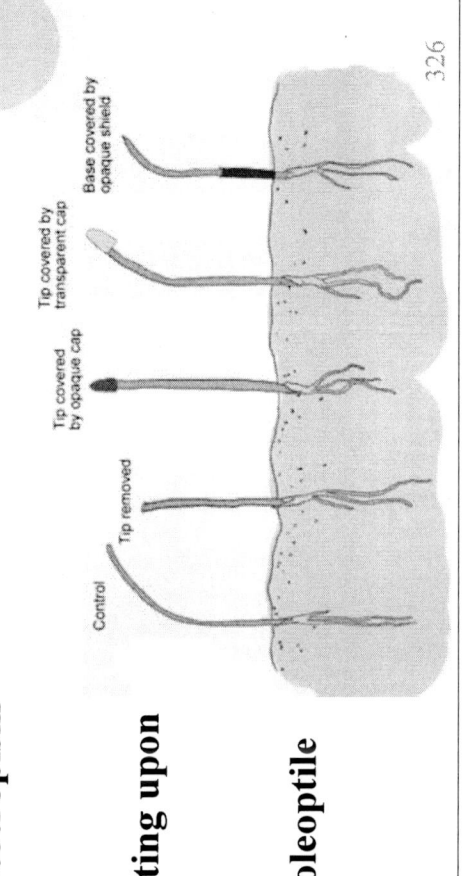

acting upon

coleoptile

Response to light

- Light cues many key events in plant growth and development

- Effects of light on plant morphology is called **photomorphogenesis**

- Plants detect not only presence of light but also its direction, intensity, and wavelength (color).

- A graph called an **action spectrum** depicts relative response of a process to different wavelengths.

Action spectra are useful in studying any process that depends on light

Absorption and action spectrum

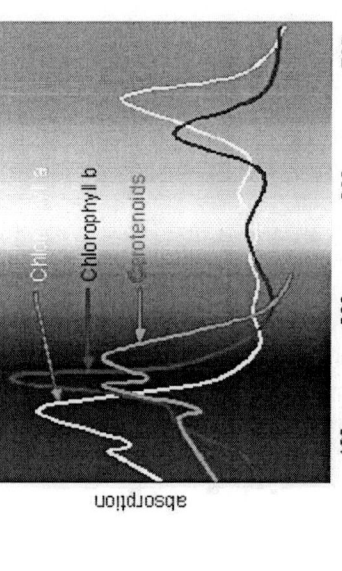

Light Receptors- Blue-Light Photoreceptors & Phytochromes

- There are 2 major classes of light receptors:

  1. Blue-light photoreceptors
  2. Phytochromes

- **1. Blue-light photoreceptors** control hypocotyl elongation, stomatal opening, and phototropism

- **2. Phytochromes** are pigments that regulate many of a plant's responses to light throughout its life.

- These responses include seed germination and shade avoidance

- Many seeds remain dormant until light conditions change

- Red light increases germination, while far-red light inhibited germination

Circadian rhythms & Plant's responses

- Many plant processes oscillate during the day

- Legumes lower their leaves in the evening & raise them in morning, even when kept under constant light or dark conditions

- **Circadian rhythms** are cycles that are about 24 hours long and are governed by an internal "clock" (day/night cycle)

- The clock may depend on synthesis of a protein regulated through feedback control and may be common to all eukaryotes

Stimuli other than light: Photoperiod

- Phytochrome conversion marks sunrise and sunset, providing the biological clock with environmental cues.

- Photoperiod, the relative lengths of night and day, is the environmental stimulus plants use most often to detect the time of year.

- **Photoperiodism** is a physiological response to photoperiod

Photoperiodism

The evening lowering of leaves by some legumes followed by their raising in the morning is a manifestation of a circadian rhythm

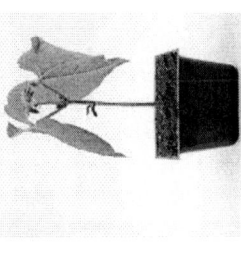

Noon   Midnight

More on Photoperiodism

- Some processes, including flowering in many species, require a certain photoperiod

- Some plants flower after only a single exposure to the required photoperiod

- Other plants need several successive days of the required photoperiod

Stimuli other than Light & Photoperiod

- Some plants need an **environmental stimulus** in addition to the required photoperiod, for example:

- **Vernalization**, which is a pretreatment with cold to induce flowering.

- Another example the flowering signal, not yet chemically identified, is called **florigen** (acting as a **hormone**)

- Florigen may be a macromolecule governed by the *CONSTANS* gene

Stimuli other than light & Photoperiod: Gravitropism

- Because of immobility, plants adjust to a range of environmental circumstances through developmental and physiological mechanisms

- One environmental circumstance that plants adjust to is gravity and this response to gravity is known as **gravitropism**

- Roots show positive gravitropism; shoots show negative gravitropism

Other stimuli: Aboitic and Biotic stresses.

- Other environmental circumstances are the aboitic and biotic stresses.

- These can adversely effect plant growth and development and plants must learn to adjust their responses and deter them.

- Stresses can be **abiotic** (nonliving) or **biotic** (living)

- **Abiotic stresses** include drought, flooding, heat stress, and cold stress, salt stress

Abiotic stresses

1. **Drought:** During drought, plants reduce transpiration by closing stomata, slowing leaf growth, & reducing exposed surface area

2. **Flooding:** Enzymatic destruction of root cortex cells creates air tubes that help plants survive oxygen deprivation during **flooding**

3. **Heat Stress:** Heat-shock proteins help protect other proteins from **heat stress** which can otherwise cause denaturation.

Abiotic stresses

4. **Cold stress:** Low temperatures decrease membrane fluidity.

- Altering lipid composition of membranes is a response to cold stress.

5. **Salt stress** where high salt concentration in soil can lower the water potential of the soil solution and reduce water uptake.

- Plants respond to salt stress by producing solutes tolerated at high concentrations

Biotic stresses

- **Herbivory,** animals eating plants, is a biotic stress that plants face in any ecosystem

- Plants use defense systems to deter herbivory and combat pathogens

- Plants counter excessive herbivory with physical defenses such as thorns & chemical defenses such as toxins

- Some plants even "recruit" predatory animals that help defend against specific herbivores

Defense against Pathogens

- A plant's first line of defense against infection is the epidermis.
- If a pathogen penetrates the dermal tissue, the second line of defense is a chemical attack that kills the pathogen and prevents its spread
- A **virulent** pathogen is one that a plant has little specific defense against.
- An **avirulent** pathogen is one that may harm but does not kill the host plant

1. Caterpillar feeds on plant & induces wounding

2. Volatile compounds are synthesized & released to attract parasitoid wasp

3. Wasps lays eggs inside caterpillar & causes its death

The 2nd line of defense in plants

1. The **hypersensitive response** causes cell and tissue death at the infection site

- It induces proteins, that attack the pathogen and stimulates changes in the cell wall to confine the pathogen

2. **Systemic acquired resistance** causes systemic expression of defense genes and is a long-lasting response

12, Origin of Animals on Earth

General characteristics of Animals

- 1.3 million living species of animals have been identified
- Animals are heterotrophs that ingest their food
- Animals are multicellular eukaryotes
- Their cells lack cell walls
- Their bodies are held together by structural proteins (collagen)
- Nervous tissue and muscle tissue are unique to animals
- Most animals reproduce sexually, with the diploid stage usually dominating the life cycle

Reproduction and Development of animals

- After a sperm fertilizes an egg, the zygote undergoes rapid mitotic divisions with no significant growth (a process known as **cleavage**)
- Cleavage leads to formation of a blastula (an embryo that has over 100 cells after Clevage has taken place)
- The blastula undergoes **gastrulation**, (cells migrating to the interior of blastula) .
- Gastrulation leads to formation of gastrula, (Embryo at this stage has different layers of embryonic tissues)

Reproduction and Development of animals

- Many animals have at least one larval stage

- **A larva** is sexually immature and morphologically distinct from the adult; it eventually undergoes metamorphosis.

- **Metamorphosis** – is the process by which an animal physically develop after birth or hatching, involving a relatively abrupt change in animal's body structure.

- All animals, and only animals, have *Hox* genes that regulate the development of body form

Great diversity of species in animal kingdom

- Although the *Hox* family of genes has been highly conserved, it can produce a wide diversity of animal morphology

- The history of animals spans more than half a billion years. The common ancestor of living animals may have lived between 675 and 875 million years ago

1. *Paleozoic Era-* (542–251 Million Years Ago)
2. *Mesozoic Era* (251–65.5 Million Years Ago)
3. *Cenozoic Era* (65.5 Million Years Ago to the Present)

Paleozoic Era (542–251 Million Years Ago)

- Within the Paleozoic era exists a Cambrian period that lasted approx. 53 million yrs. & is often known as Cambrian explosion

- The **Cambrian explosion** (million yrs. ago) marks the earliest fossil appearance of many major groups of living animals

- This period marked a dramatic burst of evolutionary changes & its cause is hypothesized in the
  - Evolving predator-prey relationships
  - Rise in atmospheric oxygen
  - The evolution of the *Hox* gene complex

Mesozoic Era (251–65.5 Million Years Ago)

- Coral reefs emerged, becoming important marine ecological niches for other organisms

- During the Mesozoic era, dinosaurs were the dominant terrestrial vertebrates

- The first mammals emerged

Cenozoic Era (65.5 Million Years Ago to the Present)

- The beginning of the Cenozoic era followed mass extinctions of both terrestrial and marine animals

- These extinctions included the large, nonflying dinosaurs and the marine reptiles

- Modern mammal orders and insects diversified during the Cenozoic era

Animals characterized by "body plans"

- Zoologists sometimes categorize animals according to a **body plan**, a set of morphological and developmental traits, integrated into an animal.

- Body Plan categorized in Morphological traits

  1. Symmetry
  2. Tissues
  3. Body Cavities

1. Symmetry

- Animals can be categorized according to the symmetry of their bodies, or lack of it

- Some animals have **radial symmetry**

- Some animals have **bilateral symmetry** which is the two-sided symmetry.

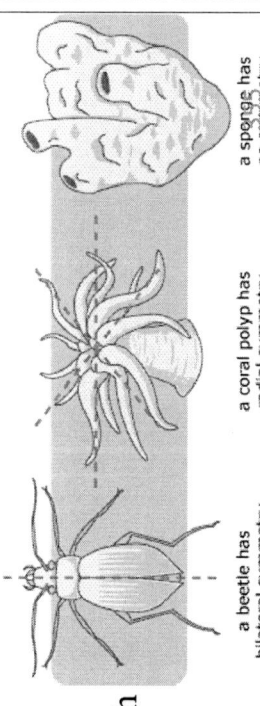

a beetle has bilateral symmetry

a coral polyp has radial symmetry

a sponge has no symmetry

2. Tissues

- Tissues are collections of specialized cells isolated from other tissues by membranous layers

- During development, three *germ layers* give rise to the tissues and organs of the animal embryo

- **Ectoderm** is the germ layer covering the embryo's surface
- **Endoderm** is the innermost germ layer
- **Mesoderm** is the middle germ layer

Diploblastic vs Triploblastic embryo

- **Diploblastic** animals have ectoderm and endoderm
- **Triploblastic** animals also have an intervening third layer called **mesoderm.**
- **Germ layers**, give rise to all of an animal's tissue & organs through organogenesis.

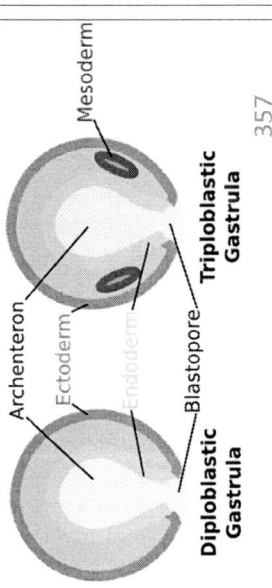

3. Body Cavities

1. **Coelomates** are animals that possess a true coelom. (Coelom is a true body cavity and is derived from mesoderm), Organs are well organized in a coelomates
2. **Pseudocoelomates** are animals that possess a pseudocoelom (a body cavity derived from mesoderm and endoderm), Organs are held loosely
3. Animals that lack a body cavity are called **Acoelomates**

13. Origin of Invertebrates

Invertebrates

Invertebrates

- **Invertebrates** are animals that neither possess nor develop a vertebral column (commonly known as a *backbone*)
- They account for 95-97% of known animal species
- There is a huge diversity among Invertebrates but in this chapter we would address only 4 kinds: Sponges, Tapeworms, Rotifers and Insects
- By far the largest number of invertebrate species are insects

Sponges

- Sponges are sedentary animals that live in both fresh and marine waters
- Sponges lack true tissues and organs
- Sponges are **suspension feeders**, capturing food particles suspended in the water that pass through their body
- Most sponges are **hermaphrodites**: Each individual functions as both male and female

Sponges

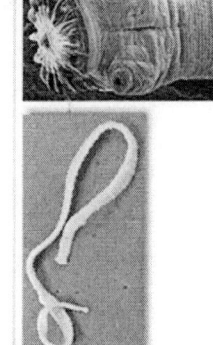
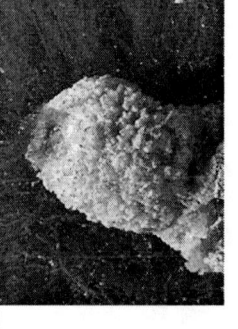
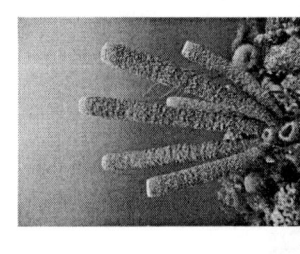

Tapeworms

- Tapeworms (adult flatworms) are parasites of vertebrates and lack a digestive system
- Live tapeworm larvae are sometimes ingested by consuming undercooked food.
- Tapeworms absorb nutrients from the host's intestine

Tapeworms life cycle

- Fertilized eggs, produced by sexual reproduction, leave the host's body in feces

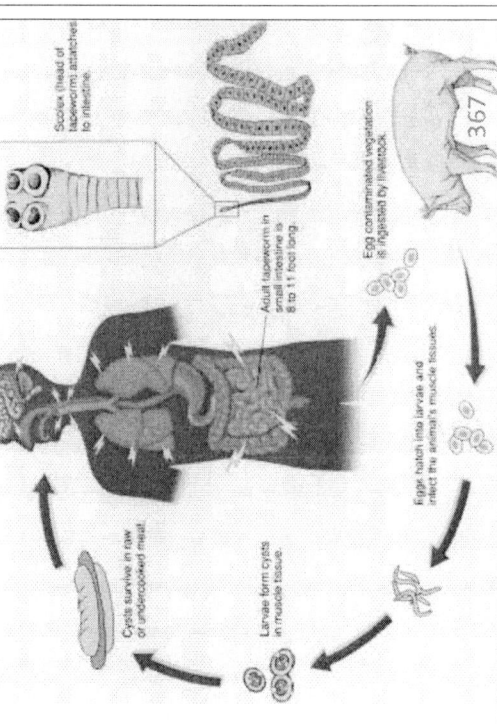

Rotifers

- Rotifers, are tiny animals that inhabit fresh water, ocean, and damp soil
- Rotifers have an **alimentary canal**, a digestive tube with a mouth & anus that lies within a fluid-filled pseudocoelom.
- Rotifers reproduce by **parthenogenesis** (an asexual reproduction) in which females produce offspring from unfertilized eggs
- Some species are unusual in that they lack males entirely

Rotifers

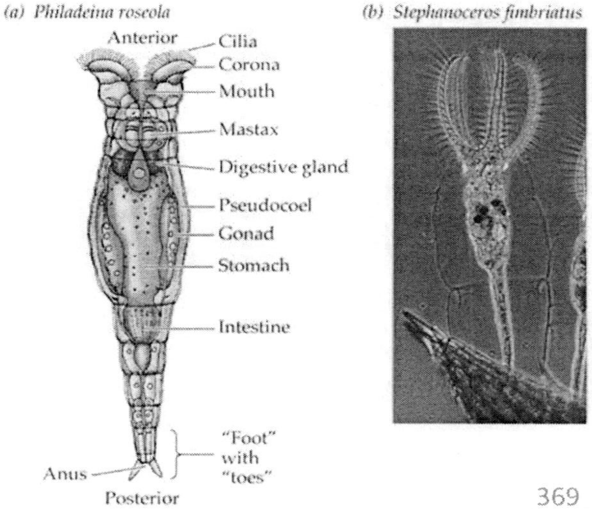
(a) Philadeina roseola
Anterior
Cilia
Corona
Mouth
Mastax
Digestive gland
Pseudocoel
Gonad
Stomach
Intestine
"Foot" with "toes"
Anus
Posterior
(b) Stephanoceros fimbriatus

369

Insects

- Insects and relatives, have more species than all other forms of life combined
- They live in almost every terrestrial habitat and in fresh water
- The anatomy of an insect includes several complex organ systems
- Flight is one key to the great success of insects
- An animal that can fly, can escape predators, find food, and disperse to new habitats much faster than organisms that can only crawl

370

Diversity of Insects

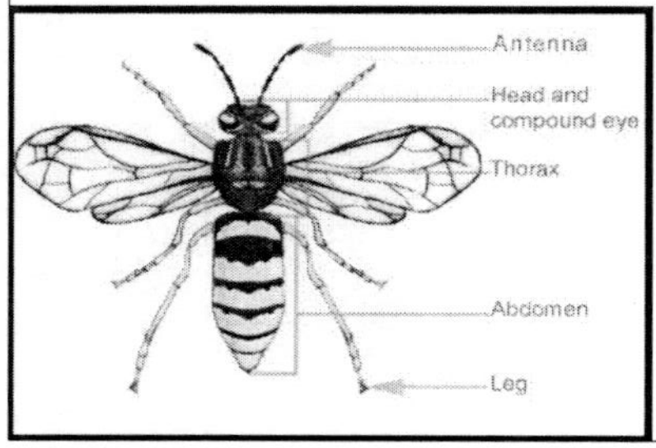
Antenna
Head and compound eye
Thorax
Abdomen
Leg

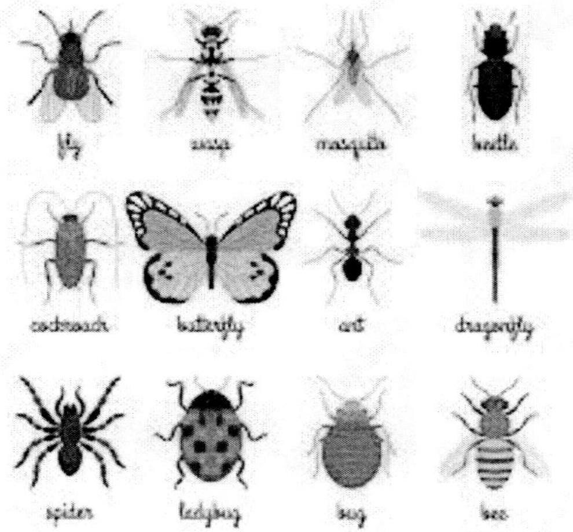
fly
wasp
mosquito
beetle
cockroach
butterfly
ant
dragonfly
spider
ladybug
bug
bee

371

- Some insects are beneficial as pollinators, while others are harmful as carriers of diseases, or pests of crops
- Many insects undergo metamorphosis during their development.
- **Metamorphosis** – is the process by which an animal physically develop after birth or hatching, involving a relatively abrupt change in animal's body structure
- **Metamorphosis** among insects is of 2 types:

1. **Complete** metamorphosis
2. **Incomplete** metamorphosis

372

Complete Metamorphosis

- Complete metamorphosis has four stages: Egg, Larva, Pupa, and Adult

- Insects with **complete metamorphosis** have larval stages and then enter an inactive state called pupa and then emerge as adults

- Insects with complete metamorphosis include beetles, bees, ants, butterflies, moths, fleas, and mosquitoes.

Incomplete Metamorphosis

- Incomplete Metamorphosis has three stages: Egg, Nymph, and Adult

- In **Incomplete metamorphosis**, the young, called nymphs, resemble adults but are smaller and lack adult features such as wing and gentialia

- Insects that have an incomplete metamorphosis life cycle include true bugs, grasshoppers, cockroaches, termites, praying mantises, crickets, and lice.

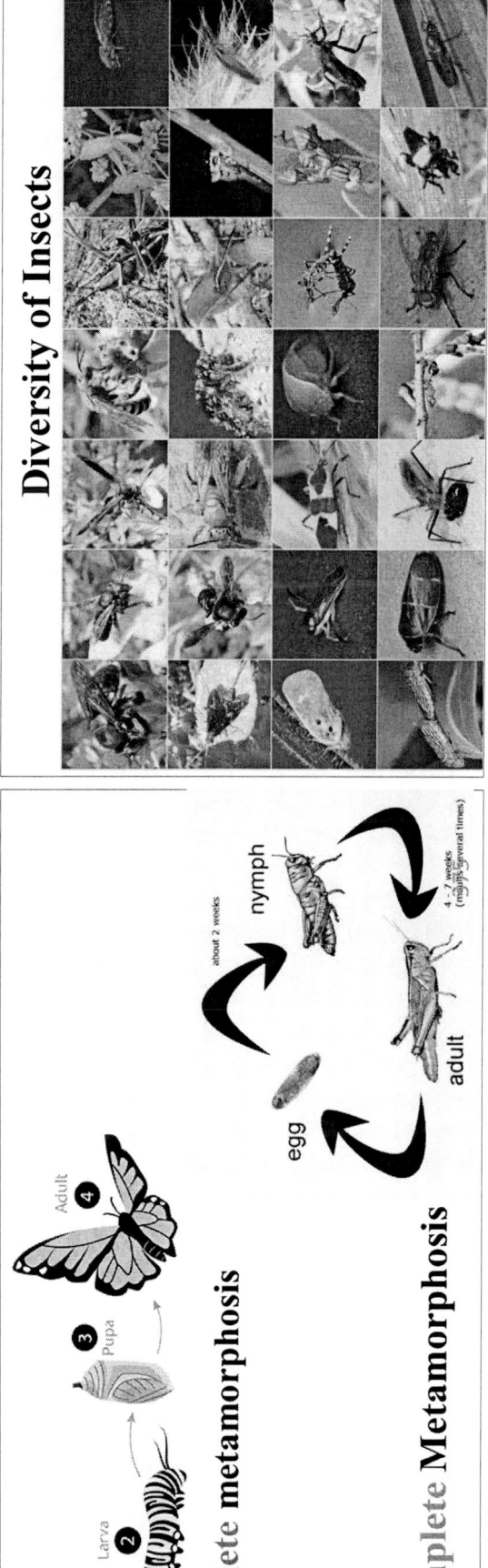

Diversity of Insects

Complete metamorphosis

Incomplete Metamorphosis

14. Origin of Vertebrates

Vertebrates- animals with backbone

- During Cambrian period, about 500 million years ago, a huge variety of animals inhabited the Earth

- One type of animal called Craniates(with head), gave rise to vertebrates(with backbone).

- The animals called **vertebrates** get their name from vertebrae, the series of bones that make up the backbone

- Vertebrates became one of the most successful groups of animal due to their complexity which made them a fierce predator

- There are over 64,000 species of vertebrates, which include the largest, fastest & strongest animals ever to inhabit the Earth

- Due to a second gene duplication involving the *Dlx gene* family, Vertebrates have the following **shared derived characters** :

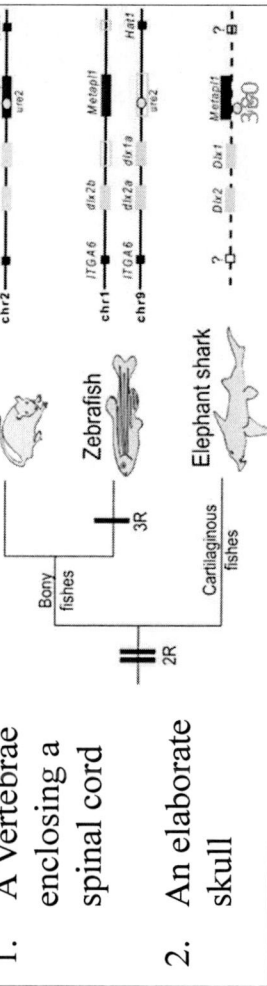

1. A Vertebrae enclosing a spinal cord
2. An elaborate skull

Gnathostomes are Vertebrates with jaws

- Shared derived Characters of Gnathostomes:
  1. Jaws
  2. An additional duplication of *Hox* genes
  3. An enlarged forebrain associated with enhanced smell and vision &
  4. In aquatic gnathostomes, **a lateral line system**, which is sensitive to vibrations

Tetrapods are Gnathostomes that have limbs

- One of the most significant events in vertebrate history was when the fins evolved into the limbs and feet of tetrapods
- **Shared** derived characters of **Tetrapods:**
  1. Four limbs, and feet with digits
  2. Ears for detecting airborne sounds

One kind of Tetrapod - Amphibians

- There are about 7500 species of **Amphibians** E.g. Frogs and Salamanders.
- *Amphibian* means "both ways of life," they start as an aquatic larva & then turn into a terrestrial adult.
- Most amphibians have moist skin that complements the lungs in gas exchange and protect them from microbes and viruses

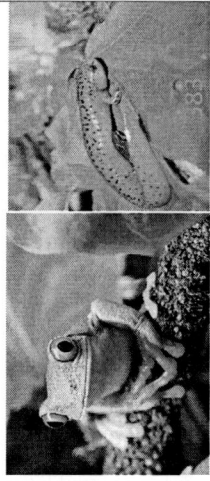

Amniotes are Tetrapods with a terrestrially adapted Egg

- **Amniotes** are a group of Tetrapods that have a terrestrially adapted egg & whose members are reptiles & mammals
- Amniotes are named for the major shared derived character the **amniotic egg**: it contains membranes that protect the embryo on a terrestrial ecosystem
- Amniotes have other terrestrial adaptations, such as relatively impermeable skin and the ability to use the rib cage to ventilate the lungs

One kind of Amniotes – Reptiles

- The **reptile** clade includes snakes, turtles, crocodiles, lizards, birds, and the extinct dinosaurs
- Reptiles have scales on their body that create waterproof barrier
- Reptiles lay their amniotic eggs on land. Innermost membrane in these eggs is called Amnion & it encloses the embryo.
- Amniotic egg also contain an amniotic fluid that surrounds the fetus and protects it from heat and physical injury.

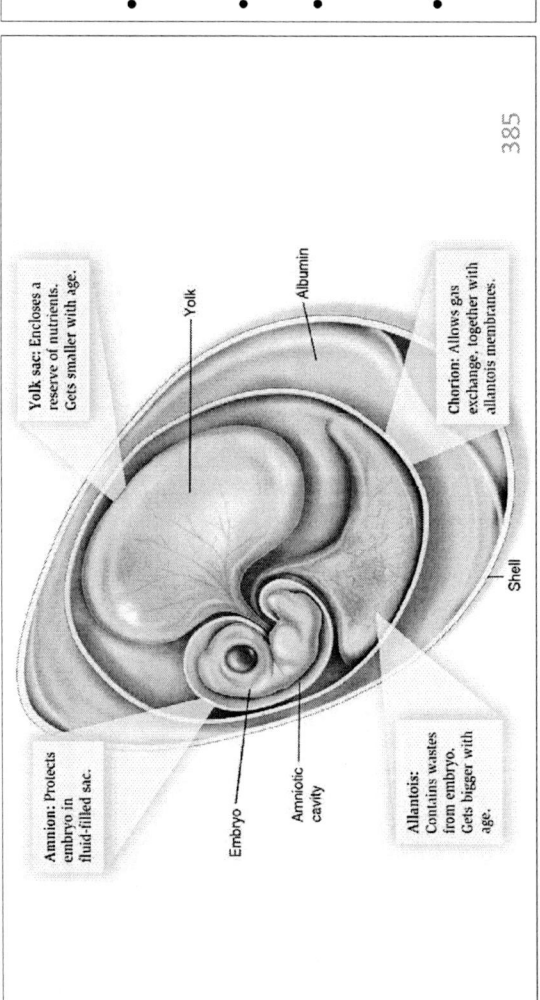

- Most reptiles are **ectothermic**, absorbing external heat as the main source of body heat, by laying out in the sun.
- Birds are **endothermic**, capable of keeping the body warm from within through metabolism

Birds (a better adapted reptile)

- The derived characters of birds are adaptations that enable flight
  1. Wings with keratin feathers
  2. Lack of a urinary bladder
  3. Females with only one ovary
  4. Small gonads, and loss of teeth
- Flight enhances hunting and scavenging, escape from terrestrial predators, and migration. However, some birds are flightless, e.g. Penguins, ducks & pigeons

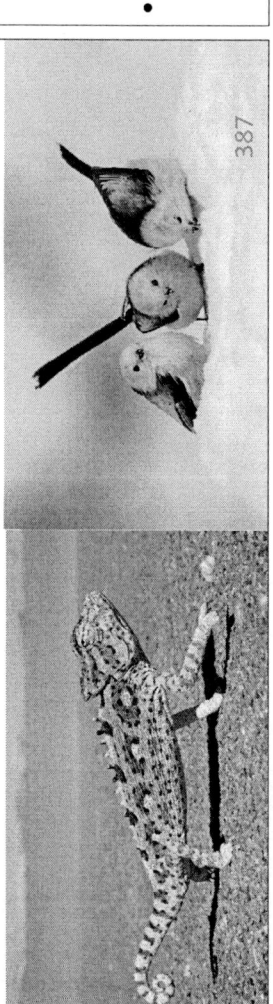

The other kind of Amniotes- Mammals

- **Mammals**, class Mammalia, consists of 5,416 species
- Mammals have the following **Derived Characters**

  —Mammary glands, which produce milk

  —Hair

  —A larger brain than other vertebrates of equivalent size

  —Differentiated teeth (capable of eating different foods)

  —Three middles ear bones (efficient and acute hearing)

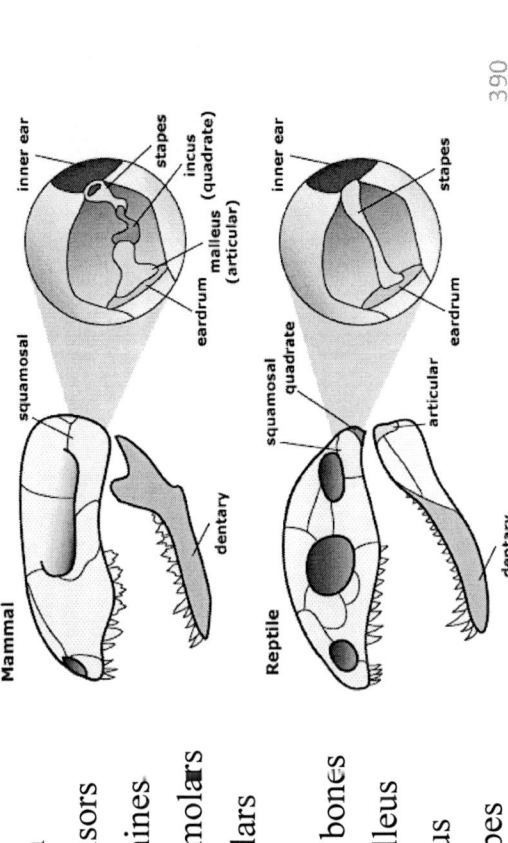

- Teeth
  1. Incisors
  2. Canines
  3. Premolars
  4. Molars
- 3 ear bones
  1. Malleus
  2. Incus
  3. Stapes

The 3 living lineages of Mammals

1. **Monotremes** - Small group of egg-laying mammals
2. **Marsupials** - Give birth to undeveloped young.
3. **Eutherians (Placentals)** Infants complete their embryonic development within uterus and are fully developed

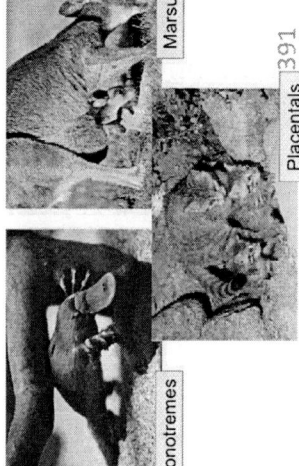

1. Monotremes

- These egg laying mammals are called Monotremes (single opening) for having one common urinary, defecatory, and reproductive duct, called the cloaca.
- The existing monotreme species are the Platypus & Echidnas.

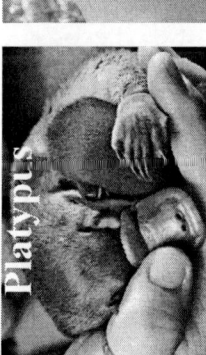

2. Marsupials

- **Marsupials** include opossums, kangaroos, and koalas
- The embryo develops within a **placenta** in the mother's uterus
- A marsupial is born very early in its development
- It completes its embryonic development while nursing in a maternal pouch called a marsupium

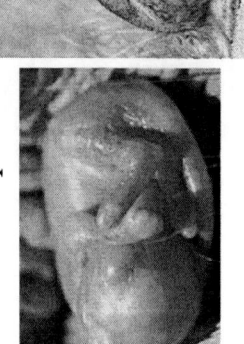

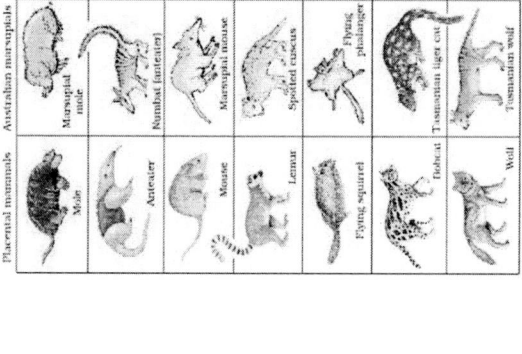

Convergent evolution has resulted in a diversity of marsupials that resemble the eutherians in other parts of the world

3. Eutherians (Placental Mammals)

- Compared with marsupials, **eutherians** have a longer gestation period
- Young eutherians complete their embryonic development within a uterus, joined to the mother by the placenta.

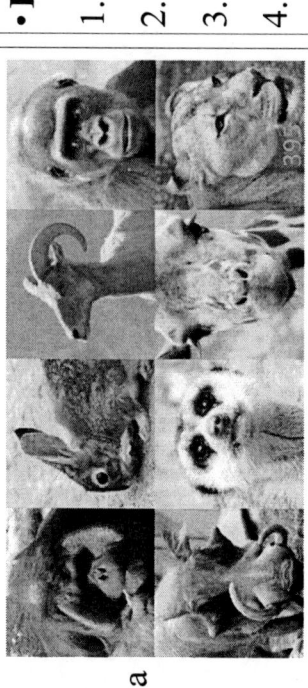

Primates belong to Eutherian Lineage

- Primates includes lemurs, monkeys, and apes. Humans are members of the ape group.
- **Derived characters of primates:**
  1. A large brain and short jaws
  2. Forward-looking eyes close together providing deep perception
  3. Complex social behavior and parental care
  4. A fully **opposable thumb** (in monkeys and apes) for grasping

Humans have a larger brain and bipedal locomotion

- The species *Homo sapiens* is about 200,000 years old, which is very young, considering life has existed for 3.5 billion years.

- **Shared derived characters** that distinguish humans from apes:
  1. Upright posture and bipedal locomotion
  2. Larger brains, language capabilities and symbolic thought
  3. The manufacture and use of complex tools
  4. Shortened jaw & Shorter digestive tract

Paleoanthropology

- Paleoanthropology is the study of human origins & has provided insight into the links between us & our primitive ancestors.

- Our primitive ancestors that we share with Chimpanzee existed around 7-8 million years ago.

- Using fossil evidence we have now learnt about a taxonomical group called Hominins (the missing links between our primitive ancestors and modern humans)

Hominins

- **Hominins** (our most recent ancestors) are more closely related to humans than to chimpanzees

- Paleoanthropologists have discovered fossils of about 20 species of extinct hominins

- Hominins originated in Africa about 6–7 million years ago

- A **common misconception** about early hominins is thinking of them as chimpanzees

Australopiths

- Australopiths were hominins living between 4 and 2 million years ago.
- Some species walked fully erect.
- The oldest evidence of tool use, cut marks on animal bones, is 2.5 million years old.
- Hominins began to walk long distances on two legs about 1.9 million years ago

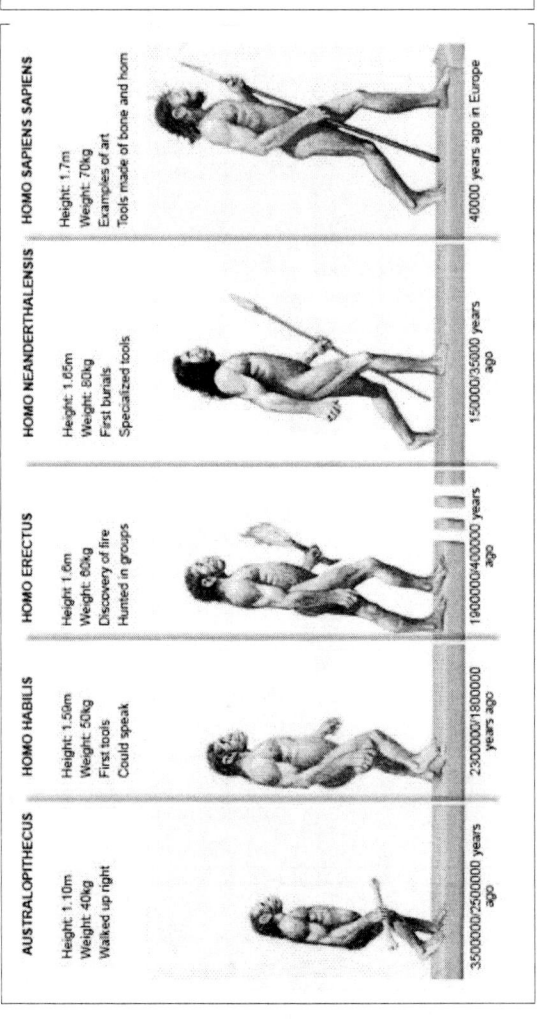

Homo ergaster

- *Homo ergaster* was the first fully bipedal, large-brained hominid
- *Homo ergaster* shows a significant decrease in sexual dimorphism (a size difference between sexes) compared with its ancestors

Homo erectus

- *Homo erectus* originated in Africa by 1.8 million years ago
- It was the first hominin to leave Africa

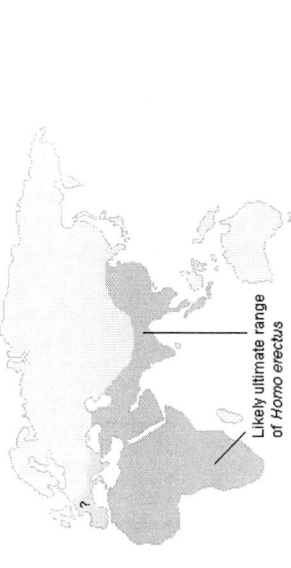
Likely ultimate range of *Homo erectus*

Neanderthals (around 200,000 to 40,000 years ago)

- Neanderthals, *Homo neanderthalensis*, were closely related to modern humans (99.7% of DNA) & lived in Europe & Asia.

- They became extinct around 40,000 years ago. They were thick-boned with a larger brain.

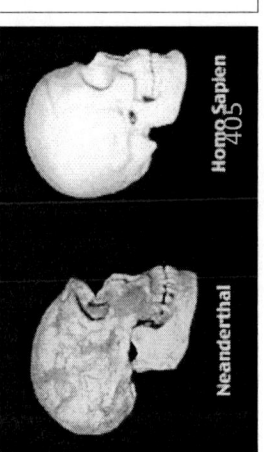

- Neanderthals contributed to the DNA of modern humans by interbreeding some 60,000 yrs. ago.

Neanderthals vs Modern Humans

- *Neanderthals* had a low surface to volume ratio with shorter legs and bigger body.

- They buried their dead & made hunting tools.

Homo Sapiens

- *Homo sapiens* appeared in Africa around 195,000 years ago & all living humans are descended from these African ancestors

- *Homo sapiens* were the first group to show evidence of symbolic and sophisticated thought

- Rapid expansion of our species may have been preceded by changes to the brain that made cognitive innovations possible. As it happened with the *FOXP2 gene* that gave us language capabilities.

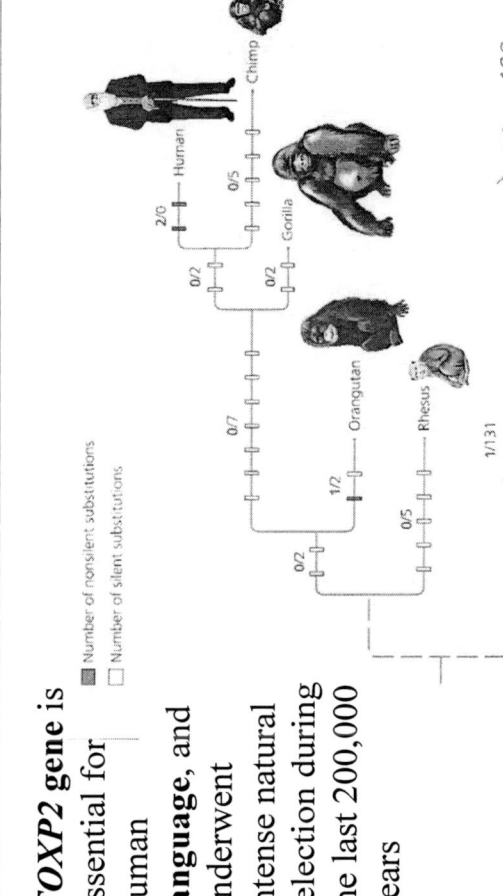

- *FOXP2 gene* is essential for human **language**, and underwent intense natural selection during the last 200,000 years

15. Anatomy & Physiology of Animals

Anatomy & Physiology : Form and Function

- **Anatomy:** Study of the structure of body parts and their relationship to one another
- **Physiology :** Study of the function of body parts; how they work to carry out life-sustaining activities
- Anatomy and physiology are inseparable
  – Function always reflects structure
  – What a structure can do depends on its specific form
  – Known as the **principle of complementarity of structure and function**

Size and shape affect the way an animal interacts with its environment

- An animal's size and shape directly affect how it exchanges energy and materials with its surroundings
- Exchange occurs as substances dissolved in the aqueous medium diffuse and are transported across the cells' plasma membranes
- In vertebrates, the space between cells is filled with **interstitial fluid**, which allows for the movement of material into and out of cells

Structural Organization

- Most animals are composed of specialized cells organized into **tissues** that have different functions
- Tissues make up **organs**, which together make up **organ systems**
- Different tissues have different structures that are suited to their functions
- Tissues are classified into four main categories: epithelial, connective, muscle, and nervous

Necessary Life Functions

- **Maintaining boundaries**: Separation between internal and external environments must exist. Plasma membranes separate cells. Skin separates organism from environment
- **Movement:** Muscular system allows movement
- **Responsiveness:** Ability to sense and respond to stimuli like a Withdrawal reflex, that prevents injury
- **Metabolism:** All chemical reactions that occur in body cells. Sum of all catabolism (breakdown of molecules) and anabolism (synthesis of molecules)

- **Digestion:** Breakdown of ingested food, followed by absorption into blood
- **Excretion**: Removal of wastes from metabolism and digestion. Urea (from breakdown of proteins), carbon dioxide (from metabolism), feces (unabsorbed foods)
- **Reproduction**: At the cellular level, reproduction involves division of cells for growth or repair. At the organismal level, reproduction is the production of offspring
- **Growth:** Increase in size of a body part or of organism

Epithelial Tissue

- **Epithelial tissue** covers the outside of the body and lines the organs and cavities within the body
- Shape of epithelial cells may be cuboidal, columnar or squamous
- The arrangement of epithelial cells may be *simple* (single cell layer), *stratified* (multiple tiers of cells), or *pseudostratified* (a single layer of cells of varying length)
- Epithelial tissue helps in protection, secretion and selective absorption

Connective Tissue

- **Connective tissue** mainly binds and supports other tissues
- It contains sparsely packed cells scattered throughout an extracellular matrix
- There are 3 types of connective tissue fiber, all made of protein
  1. *Collagenous fibers* provide strength and flexibility
  2. *Elastic fibers* stretch and snap back to their original length
  3. *Reticular fibers* join connective tissue to adjacent tissues

Six major types of connective tissue

1. **Loose** connective tissue binds epithelia to underlying tissues and holds organs in place
2. **Cartilage** is a strong and flexible support material
3. **Fibrous** connective tissue is found in tendons, which attach muscles to bones, & ligaments, which connect bones at joints
4. **Adipose** tissue stores fat for insulation and fuel
5. **Blood** is composed of blood cells and plasma
6. **Bone** is mineralized and forms the skeleton

Muscle Tissue

- **Muscle tissue** consists of long cells called muscle fibers, which contract in response to nerve signals
- It is divided in the vertebrate body into three types:
  1. **Skeletal muscle**, or striated muscle, is responsible for voluntary movement
  2. **Smooth muscle** is responsible for involuntary body activities
  3. **Cardiac muscle** is responsible for contraction of the heart

Nervous Tissue

- **Nervous tissue** senses stimuli and transmits signals throughout the animal
- Nervous tissue contains:
  – **Neurons**, or nerve cells, that transmit nerve impulses
  – **Glial cells**, or **glia**, that help nourish, insulate, and replenish neurons

Control and Coordination – depends on the **endocrine** and **nervous system**

1. *The endocrine system* transmits chemical signals called **hormones** to receptive cells throughout the body via blood
   - Hormones are slow acting, but can have long-lasting effects
2. *The nervous system* transmits information between specific locations
   - Nerve signal transmission is very fast. Nerve impulses can be received by neurons, muscle cells, and endocrine cells

Nervous & Endocrine Systems

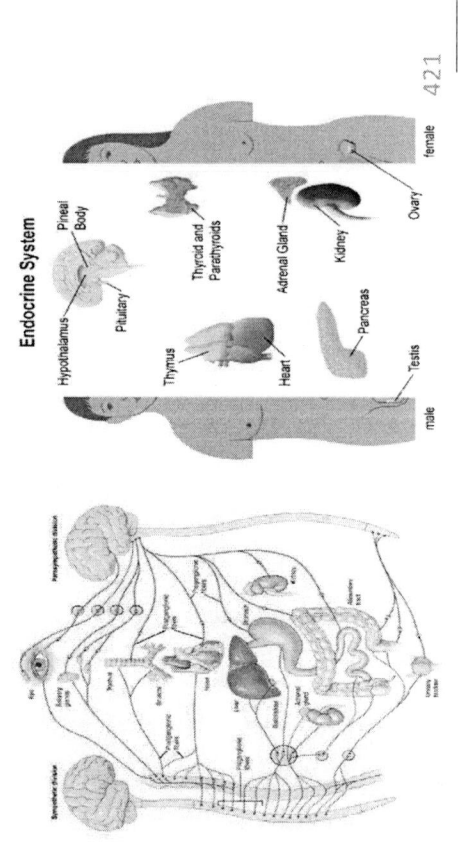

Feedback loops maintain the internal environment

- Animals manage their internal environment by regulating or conforming to the external environment
- A **regulator** uses internal mechanisms to moderate internal change in the face of an external, environmental fluctuation
- A **conformer** allows its internal condition to vary with certain external changes
- This management of internal environment is the process we know as **Homeostasis**

Regulator vs Conformer

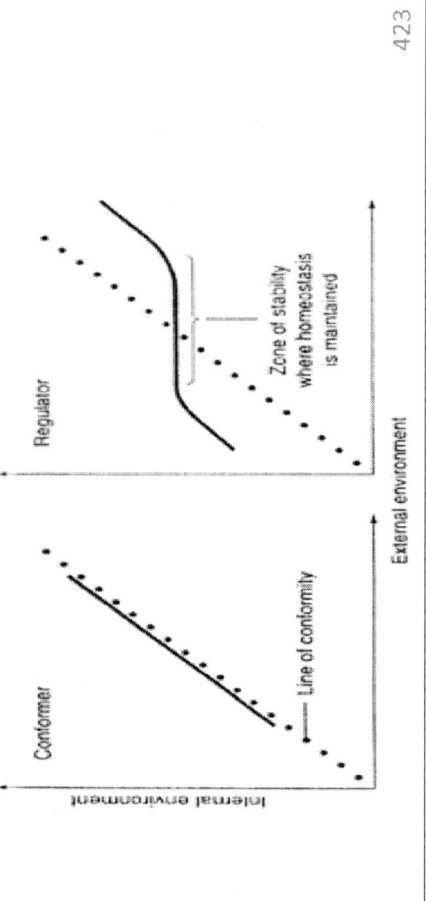

Homeostasis

- Organisms use **homeostasis** to maintain a "steady state" or internal balance regardless of external environment
- In humans, **variables** such as body temperature, blood pH, & glucose concentration are each maintained at a constant level
- For a given variable, fluctuations above or below a **set point** serve as a **stimulus**; these are detected by a **sensor**, which trigger a **response**. The response then returns the variable to the set point.

Homeostasis & Negative feedback loop

- Mechanisms of homeostasis moderate changes in the internal environment
- Homeostasis is maintained by **negative feedback**, which helps to return a variable to either a **normal range** or a set point
- Most homeostatic control systems function by negative feedback, where buildup of the end product shuts the system off
- Homeostasis can adjust to changes in external environment, a process called **acclimatization**

Homeostatic control of variables involve 3 components receptor, control center, and effector

- **Receptor** (sensor): Monitors environment & responds to **stimuli** (things that cause changes in controlled variables)
- **Control center:** Determines set point at which variable is maintained. Receives input from receptor & determines appropriate response
- **Effector:** Receives output from control center. Response either reduces stimulus (− feedback) or enhances stimulus (+ feedback)

Positive feedback loop

- **Positive feedback:** Response enhances or exaggerates the original stimulus
- Usually controls infrequent events that do not require continuous adjustment, for example:
  - Enhancement of labor contractions by oxytocin
  - Platelet plug formation and blood clotting

Negative feedback loop

- Response reduces or shuts off original stimulus. Examples:
  - ➢ Regulation of body temperature (nervous system mechanism)
  - ➢ Regulation of blood glucose (endocrine system mechanism)
- The blood glucose (sugar) is regulated by Insulin, when the **receptors** sense increased blood glucose & Pancreas (**control center**) secretes insulin into the blood
- Insulin causes body cells (**effectors**) to absorb more glucose, which decreases blood glucose levels

Homeostatic processes for thermoregulation

- **Thermoregulation** is the process by which animals maintain an internal temperature within a tolerable range

- **Endothermic** animals generate heat by metabolism; birds and mammals are endotherms

- **Ectothermic** animals gain heat from external sources; ectotherms include most invertebrates, fishes, amphibians, and non-avian reptiles

- Endothermy is more energetically expensive than ectothermy

Examples of some Endotherms & Ectotherms

Exchanging heat with the environment

- Organisms **exchange** heat by 4 physical processes: conduction, convection, radiation, and evaporation

- Heat **regulation** (thermoregulation) in mammals often involves the **integumentary system**: skin, hair, and nails

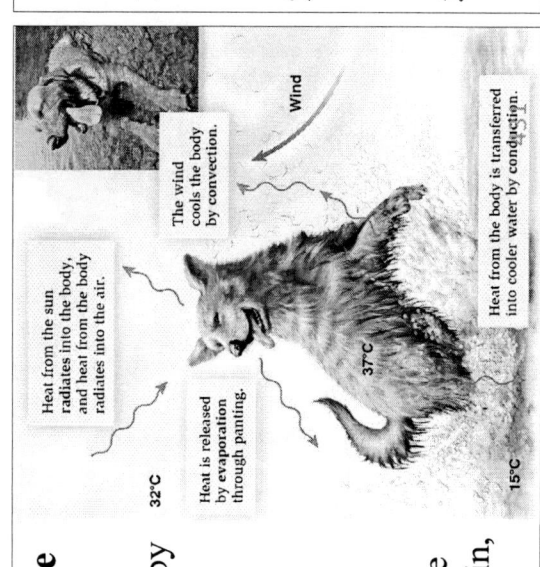

Five adaptations that help animals Thermoregulate

1. **Insulation**- Insulation is a major thermoregulatory adaptation in mammals and birds. Skin, feathers, fur, and blubber reduce heat flow between an animal and its environment

2. **Cooling by evaporative heat loss**- Many types of animals lose heat through evaporation of water in sweat. Sweating or bathing moistens the skin, helping to cool an animal down

3. **Behavioral responses**- Some terrestrial invertebrates have postures that minimize or maximize absorption of solar heat

4. **Adjusting metabolic heat production** - Some animals adjust their rate of metabolic heat production by increasing their muscle activity such as moving or shivering, that generates heat.

5. **Circulatory Adaptations**: Regulation of blood flow near the body surface significantly affects thermoregulation

a) In **vasodilation**, blood flow in the skin increases, facilitating heat loss

b) In **vasoconstriction**, blood flow in the skin decreases, lowering heat loss

Acclimatization in Thermoregulation

- Endotherms (Birds and mammals) can vary their insulation to acclimatize to seasonal temperature changes

- When temperatures are subzero, some ectotherms produce "antifreeze" compounds to prevent ice formation in their cells

- Thermoregulation is controlled by a region of the brain called the **hypothalamus**

- The hypothalamus triggers heat loss or heat generating mechanisms

Homeostasis: Internal temperature of 36–38°C

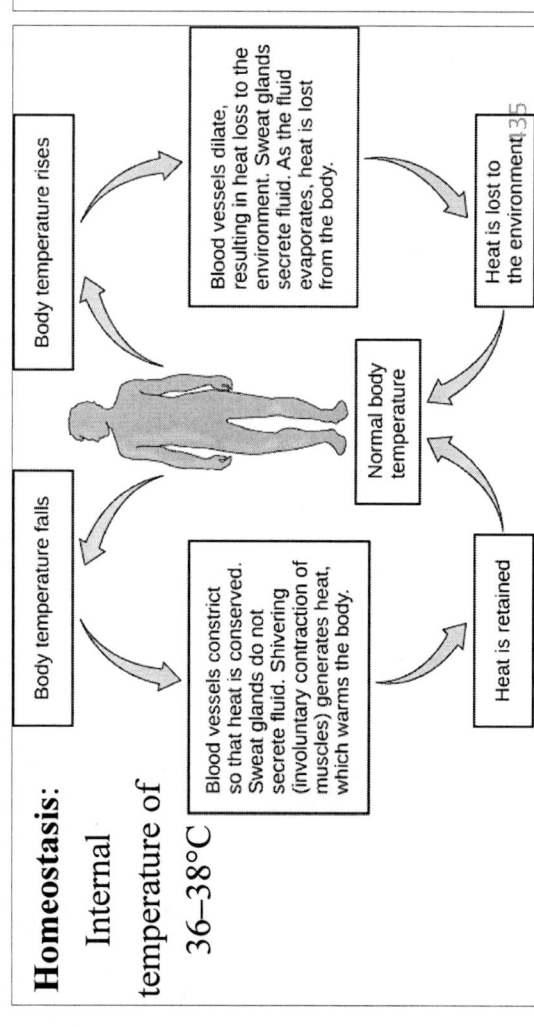

Bioenergetics

- **Bioenergetics** is the overall flow and transformation of energy in an animal

- It is concerned with the energy involved in making & breaking of chemical bonds in the molecules found in the organisms

- Energy molecules in food are broken (catabolism) to make ATP, which powers cellular work (anabolism)

- **Bioenergetics** determines how much food an animal needs and relates to an animal's size, activity, and environment

BMR & SMR

- **Metabolic rate** is the amount of energy an animal uses in a unit of time

- One way to measure it is to determine the amount of oxygen consumed & carbon dioxide produced

- **Basal metabolic rate (BMR)** is the metabolic rate of an endotherm at rest at a "comfortable" temperature

- **Standard metabolic rate (SMR)** is the metabolic rate of an ectotherm at rest at a specific temperature

Metabolic rate

- Both BMR and SMR assume a nongrowing, fasting, and nonstressed animal.

- Ectotherms have much lower metabolic rates than endotherms of a comparable size

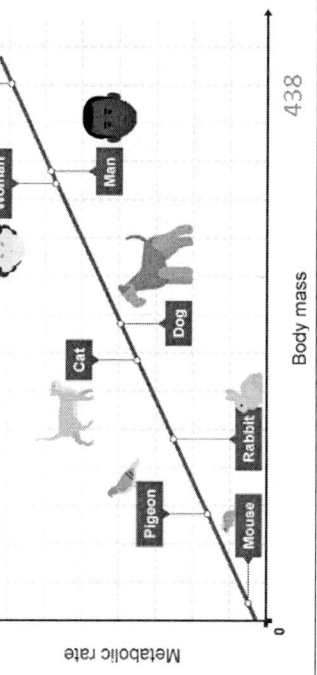

Torpor for unfavorable conditions

- **Torpor** is a physiological state, an adaptation, in which activity is low and metabolism decreases

- Torpor enables animals to save energy while avoiding difficult and dangerous conditions

- **Hibernation** is long-term torpor that is an adaptation to winter cold and food scarcity

- **Estivation**, or summer torpor, enables animals to survive long periods of high temperatures and scarce water supplies

- **Hibernation** helps to avoid damage from low temperatures in winter.

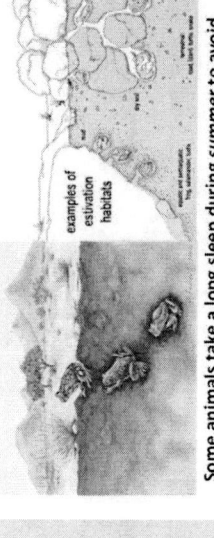

- **Estivation** helps to avoid damage from high temperatures in summer.

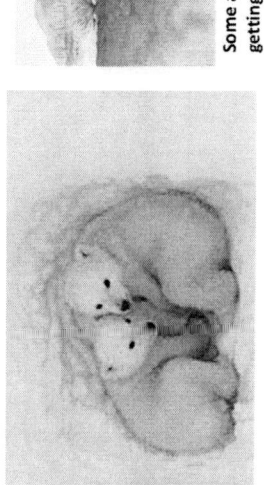

Some animals take a long sleep during *summer* to avoid getting dried up

16, Animal Nutrition

Animal nutrition

- Food is taken in, taken apart, and taken up in the process of animal nutrition
- In general, animals fall into three categories:
  1. **Herbivores** eat mainly autotrophs (plants and algae)
  2. **Carnivores** eat other animals
  3. **Omnivores** regularly consume animals as well as plants

USDA's My Plate guidelines

- Portions on a dinner plate
  - **Fruits**
  - **Vegetables**
  - **Grains**
  - **Protein**
  - **Dairy**
- Basic dietary principles: eat only what you need (eat less overall); eat plenty of fruits, vegetables, and whole grains; avoid junk food

Nutrients

- **Nutrient**: substance in food needed for growth, maintenance, repair.
- Most nutrients are used for metabolic fuel, but some are for cell structures and molecular synthesis
- **Macronutrients**: three major nutrients that make up the bulk of ingested food: Carbohydrates, lipids, and proteins
- **Micronutrients**: two other nutrients that are required, but only in small amounts are Vitamins and minerals

The role of Carbohydrates

- Used as fuel by cells to generate ATP during Respiration. Example: Glucose
- Some cells use fat for energy but Neurons and RBCs rely entirely on glucose. Neurons die quickly without glucose
- Excess glucose is converted to glycogen or fat, then stored
- Fructose and galactose are converted to glucose by liver before entering circulation

The role of Lipids

- The Adipose tissue offers protection, insulation, fuel storage
- Phospholipids essential in myelin sheaths and all cell membranes
- Cholesterol stabilizes membranes
- Prostaglandins → smooth muscle contraction, BP control, inflammation
- Help absorb fat-soluble vitamins

The role of Proteins

- Make up the building blocks, or structural materials
  - Example: keratin (skin), collagen and elastin (connective tissue), and muscle proteins
- Plays a vital role in several body functions as functional molecules
  - Example: enzymes and some hormones

An animal's diet must supply Chemical energy, Organic molecules & Essential nutrients

- An animal's diet provides chemical energy, which is converted into ATP and powers processes in the body
- Animals also need a source of organic carbon and organic nitrogen in order to construct organic molecules
- **Essential nutrients** are required by cells and must be obtained from dietary sources

Essential Nutrients

- Must contain both macronutrients and micro nutrients

- Essential nutrients are required by cells and must be obtained from dietary sources. There are 4 classes of essential nutrients:

    1. Essential amino acids
    2. Essential fatty acids
    3. Vitamins
    4. Minerals

Essential Amino Acids

- Animals require 20 amino acids and can synthesize about half from molecules in their diet

- The remaining amino acids, the **essential amino acids**, must be obtained from food in preassembled form

- A diet that provides insufficient essential amino acids causes malnutrition called protein deficiency

- Meat, eggs, and cheese provide all the essential amino acids and are thus "complete" proteins

- Individuals who eat only plant proteins need to eat specific plant combinations to get all essential amino acids

Essential Fatty Acids

- Animals can synthesize most of the fatty acids they need
- The **essential fatty acids** are certain unsaturated fatty acids that must be obtained from diet. Deficiencies in fatty acids are rare
- Liver can convert some fatty acids into others, but essential fatty acids (examples: linoleic and linolenic acid found in most oils) must be eaten.
- Essential fatty acids make up lipids, that have a very wide range of applications within the human body

Vitamins

- Are Organic compounds that are crucial in helping body use nutrients.
- Most function as **coenzymes.**
- Most must be ingested, except: Vitamin D (made in skin). Some B and K synthesized by intestinal bacteria. Beta-carotene converted in body to vitamin A.
- 13 vitamins essential to humans have been identified

Vitamins

- No one food group contains all vitamins.
- Two types of vitamins based on solubility
  1. **Water-soluble** vitamins: B complex and C
  2. **Fat-soluble** vitamins A, D, E, and K
- Vitamins C, E, and A and mineral selenium are *antioxidants* that neutralize these free radicals.
- Found in Broccoli, Cauliflower, Brussels sprouts.

Minerals

- **Minerals** are simple inorganic nutrients, usually required in small amounts
- Seven **minerals** are required in moderate amounts: Calcium, phosphorus, potassium, sulfur, sodium, chlorine, and magnesium
- Others are required in trace amounts
- Mineral-rich foods: Vegetables, legumes, milk, some meats

- Uptake and excretion of Minerals are balanced to prevent toxic overload.

- Examples of uses in body:

  - Calcium, phosphorus, and magnesium salts harden bone
  - Iron is essential for oxygen binding to hemoglobin
  - Iodine is necessary for thyroid hormone synthesis
  - Sodium and chloride are major electrolytes in blood

Dietary Deficiencies

- Insights into human nutrition have come from *epidemiology*, the study of human health and disease in populations

- **Undernourishment** is the result of a diet that consistently supplies less chemical energy than the body requires

- **Malnourishment** is the long-term absence from the diet of one or more essential nutrients

- An undernourished individual will
  - Use up stored fat and carbohydrates
  - Break down its own proteins and lose muscle mass
  - Suffer protein deficiency of the brain
  - Die or suffer irreversible damage

- Malnourishment can cause deformities, disease, and death
- Malnourishment can be corrected by changes to a diet

Processing of food involves 6 essential activities:

1. **Ingestion** is the act of eating and is of 4 kinds

   a. **Suspension Feeders**- Many aquatic animals are suspension feeders, which sift food particles from the water. E.g Whale

   b. **Substrate Feeders**- These are animals that live in or on their food source. E.g.: Caterpillar

   c. **Fluid Feeders**- are those that suck nutrient-rich fluid from a living host. E.g. Mosquito

   d. **Bulk feeders** eat relatively large pieces of food e.g. human, Python

2. **Propulsion**: movement of food through the alimentary canal, which includes: Swallowing & Peristalsis

- **Peristalsis**: major means of propulsion of food that involves alternating waves of contraction and relaxation.

3. **Mechanical breakdown**: includes chewing, mixing food with saliva, churning food in stomach, and **segmentation**

- Segmentation: local constriction of intestine that mixes food with digestive juices

461

4. **Digestion**: series of catabolic steps that involves enzymes that break down complex food molecules into chemical building blocks.

- In chemical digestion, the process of **enzymatic hydrolysis** splits bonds in molecules with the addition of water

5. **Absorption**: passage of digested fragments from lumen of GI tract into blood or lymph

6. **Defecation**: elimination of indigestible substances via anus in form of feces.

462

Digestive Compartments

- Most animals process food in specialized compartments. These compartments reduce the risk of an animal digesting its own cells & tissues

a) **Intracellular digestion**, the food particles are engulfed by endocytosis and digested within food vacuoles

b) **Extracellular digestion** is the breakdown of food particles outside of cells

463

Two kinds of Extracellular digestion

1. Animals with simple body plans have a **gastrovascular cavity** that functions in both digestion and distribution of nutrients

2. More complex animals have a digestive tube with two openings, a mouth and an anus. This digestive tube is called a **complete digestive tract** or an **alimentary canal**

- It can have specialized regions that carry out digestion and absorption in a stepwise fashion

464

Specialized Organs for food processing in the mammalian digestive system

- The mammalian digestive system consists of an alimentary canal and accessory glands that secrete digestive juices through ducts

- Mammalian accessory glands are the salivary glands, the pancreas, the liver, and the gallbladder

- Food is pushed along by **peristalsis**, rhythmic contractions of muscles in the wall of the canal

The Digestive system

- Valves called **sphincters** regulate the movement of material between compartments

The Oral Cavity, Pharynx, and Esophagus

- The first stage of digestion is mechanical and takes place in the **oral cavity.**

- Then **Salivary glands** deliver saliva to lubricate food & the enzyme salivary **amylase** initiates breakdown of glucose polymers.

- The tongue shapes food into a **bolus** & provides help with swallowing. Then **esophagus** conducts food from the pharynx down to the stomach by peristalsis

Digestion in the Stomach

- The **stomach** stores food and secretes **gastric juice**, which converts a meal to acid **chime**. Gastric juice is made up of HCL and the enzyme **pepsin.**

- The 3 main types of gland cells in the lining of the stomach

1. *Parietal cells:* secrete hydrogen & chloride ions separately.
2. *Chief cells:* secrete inactive **pepsinogen**, which is activated to pepsin when mixed with HCL in the stomach.
3. *Mucin cells*: secrete Mucus

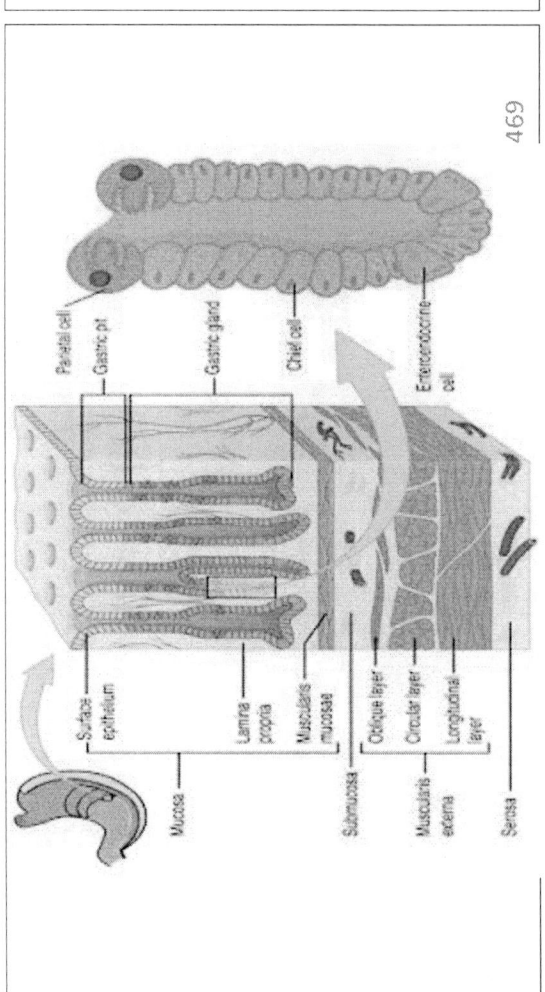

- Sphincters prevent chyme from entering the esophagus and regulate its entry into the small intestine.

- Stomach denatures proteins with the help of HCL and uses Pepsin to carry out enzymatic digestion of proteins.

- Another stomach function essential to life is secretion of **intrinsic factor (by Parietal cells)** for absorption of vitamin B_{12} in small intestine.

- Vitamin B_{12} is needed for red blood cells to mature.

Role of Pancreas

- The **pancreas** produces proteases trypsin and chymotrypsin, protein-digesting enzymes that are activated after entering the duodenum

- Its solution is alkaline and neutralizes the acidic chyme

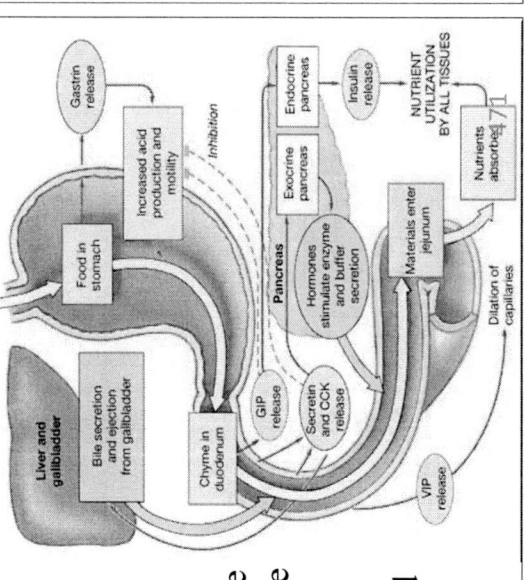

Role of Liver & Gallbladder

- **Liver**'s digestive function is production of bile which functions as a fat emulsifier, while the gallbladder stores excess bile.

- In the small intestine, **bile** aids in digestion & absorption of fats.

- **Bile appears as** Yellow-green, alkaline solution that contains :

- **Bile salts** (cholesterol derivatives) & **Bilirubin**: pigment formed from heme, after breakdown of old RBC's

Digestion in the Small Intestine

- The **small intestine** is the longest section of the alimentary canal & has 3 subdivisions:
- **Duodenum:** (10 inch), **Jejunum:** (8 ft), **Ileum:** (12 ft)
- Small Intestine is the major organ of digestion & absorption
- The first portion of the small intestine is the **duodenum**, where acid chyme from the stomach mixes with digestive juices from the pancreas, liver, gallbladder, and the small intestine itself

Secretions and absorptions in the Small Intestine

- The epithelial lining of the duodenum, called the brush border, produces several digestive enzymes
- Enzymatic digestion is completed as peristalsis moves the chyme and digestive juices along the small intestine
- The small intestine has a huge surface area, due to **villi** and **microvilli** (cellular membrane protrusions exposed to intestinal lumen) which greatly increases the rate of nutrient absorption

Structure of the small intestine

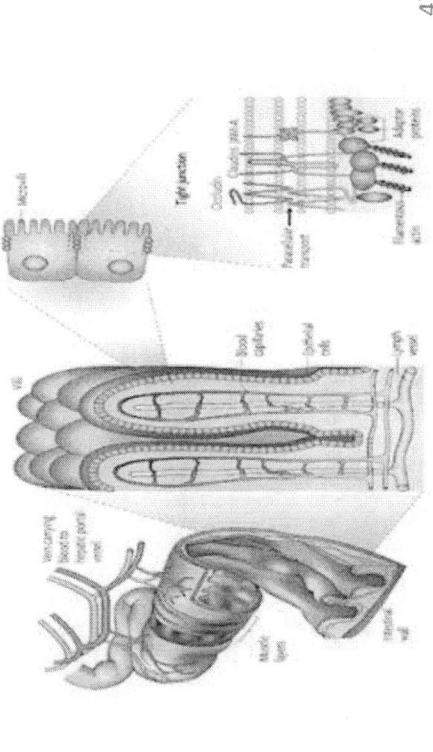

Transport of Amino acids, sugars & Fats via Small intestine

- Amino acics and sugars pass through the epithelium of the small intestine and directly enter the blood stream by diffusion
- Fats are transported by the capillaries and veins from the lacteals(small lymphatic vessels in the villi) which converge in the **hepatic portal vein.**
- **HPV** then delivers nutrient rich blood to the liver and then on to the heart.

Large intestine

- **Subdivisions of large intestine**

1. **Cecum:** first part of large intestine

2. **Appendix:** masses of lymphoid tissue & bacterial storehouse capable of recolonizing gut when necessary

3. **Colon:** has several region: Ascending colon, Transverse colon, Descending colon, Sigmoid colon

4. **Rectum:** Feces pass through the rectum and exit via the anus

- The **cecum** aids in the fermentation of plant material and connects where the small and large intestines meet

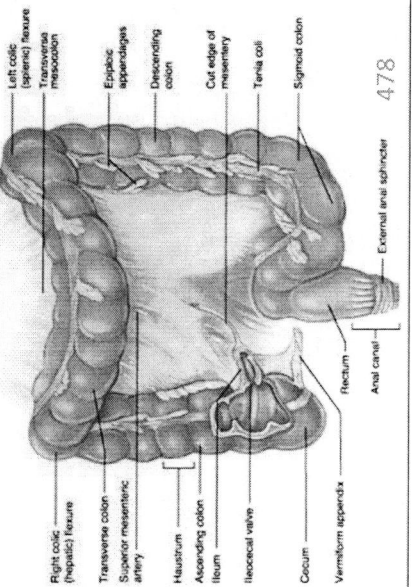

- The human cecum has an extension called the **appendix**, which plays a very minor role in immunity

- The **colon of large intestine** is connected to the small intestine

Absorption in the Large Intestine

- A major function of the colon is to recover Vitamin, water, and electrolytes (especially Na^+ & Cl^-)

- Bacterial flora: consist of 1000+ different types of bacteria

- **Metabolic functions of Bacterial flora**

- Ferments indigestible carbohydrates and mucin
- Synthesize B complex and some vitamin K needed by liver to produce clotting factors
- Beneficial bacteria outnumber and suppress pathogenic bacteria

Adaptations of vertebrate digestive systems

- Many herbivores have fermentation chambers, where symbiotic microorganisms digest cellulose

- The most elaborate adaptations for a herbivorous diet have evolved in the animals called **ruminants**

- These are mammals that acquire nutrients from plant-based food by fermenting it in a specialized stomach prior to digestion

- Instead of 1 compartment to the stomach they have 4, of which rumen is the largest filled with billions of tiny microorganisms that are able to break down grass and other coarse vegetation

Herbivores have longer cecum

- Herbivores generally have longer alimentary canals than carnivores, reflecting the longer time needed to digest vegetation

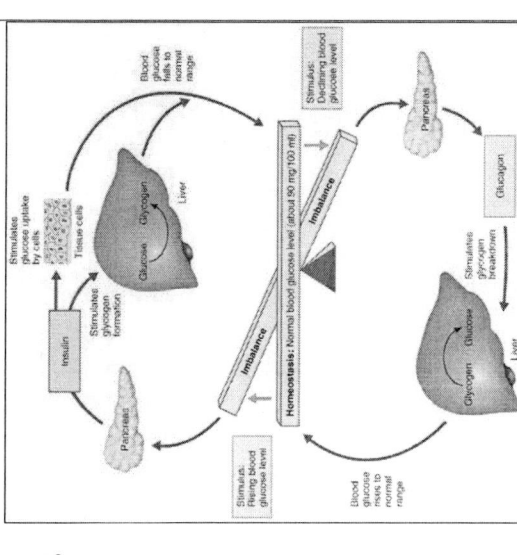

Homeostatic mechanisms and balance of energy

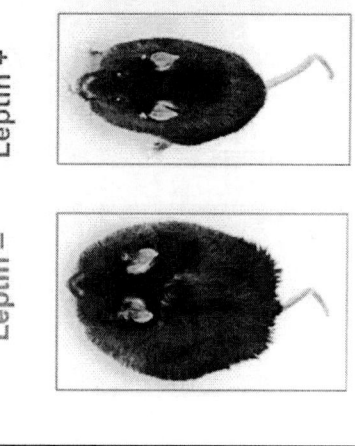

- Animals store excess calories primarily as glycogen in the liver and muscles and as fat in adipose tissue.

Overnourishment and Obesity

- **Overnourishment** causes obesity, which results from excessive intake of food energy with the excess stored as fat

- Obesity contributes to diabetes (type 2), cancer of the colon and breasts, heart attacks, and strokes

- Hormones like Leptin, Ghrelin, PYY, and Insulin regulate long-term and short-term appetite by affecting a "satiety center" in the brain

The role of Leptin

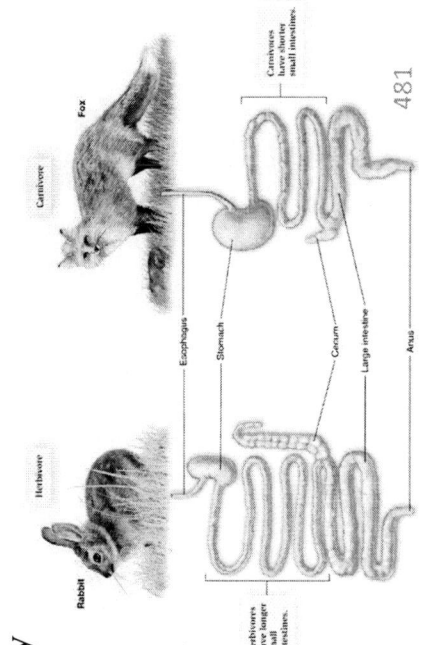

Leptin −   Leptin +

- The complexity of weight control in humans is evident from studies of the hormone Leptin

- Mice that inherit a defect in the gene for leptin become very obese

17. Circulatory & Respiratory System

Introduction

- All organisms exchange materials with their environment and these exchanges occur at the cellular level

- Unlike unicellular organisms, multicellular organisms can't direct exchange material with the environment.

- Hence multicellular organisms need specialized structures and organs(like Lungs & Gills) to help them achieve that.

Open and Closed Circulatory systems

- Complex animals have either open or closed circulatory systems that circulate fluid.

- Both open and closed systems have three basic components:

  1. A circulatory fluid (blood or hemolymph)
  2. A set of tubes (blood vessels)
  3. A muscular pump (the **heart**)

Open Circulatory system

- Arthropods (insects & others) and Molluscs, have an **open circulatory system** where the blood is in direct contact with the organs (no arteries, veins, capillaries involved)

- In an open circulatory system, blood mixes with the interstitial fluid(tissue fluid), and this combined body fluid is called **hemolymph**.

- The interstitial fluid is the fluid found in the extracellular spaces (between cells). It is part of ECM or extra cellular matrix.

Closed Circulatory system

- In a **closed circulatory system**, blood is confined to vessels and is separated from the interstitial fluid.

- The plasma part of the blood is what ends up becoming the Interstitial fluid when it diffuses out of the capillaries and into the ECM

- Closed systems are found in complex animals, since these systems are more efficient at transporting fluids to tissues and cells, where an open system wont reach.

Open Vs Closed Circulatory System

Common to both

1. Circulatory fluid
2. Set of tubes
3. Muscular pump

Closed Circulatory system

- Humans and other vertebrates have a closed circulatory system, called the **cardiovascular system**

- Vertebrate hearts contain two or more chambers where blood enters through **atrium** & is pumped out through **ventricle**

- Two types of closed circulatory system
1. Single circulation
2. Double circulation

Single circulation

- Single circulation occurs in a 2 chambered heart and is found in bony fishes, rays, and sharks

- In single circulation, blood leaving the heart passes through two capillary beds before returning

Double Circulation

- Double circulation occurs in a 3 and 4 chambered heart and is found in Amphibian, Reptiles, and Mammals.

- In reptiles and mammals, oxygen-poor blood flows through the **pulmonary circuit** to pick up oxygen through the lungs

- In **Amphibians through Pulmocutaneous circuit** to pick up oxygen through the lungs and skin.

Double Circulation

- **Mammals and birds** have a four-chambered heart with two atria and two ventricles

- Mammals and birds are endotherms and require more O_2 than ectotherms

- In double circulation, oxygen-poor & rich blood are pumped separately by the heart into pulmonary and systemic circuits respectively.

- Double circulation maintains higher blood pressure in the organs than does single circulation

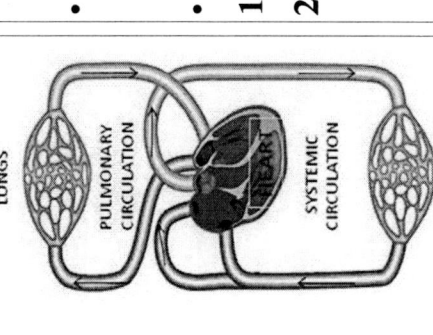

Cardiovascular system

- The mammalian circulatory systems are much more complex, an example of such a system is the human **cardiovascular system**

- The 3 main types of blood vessels in a cardiovascular system are

1. **Arteries:** They carry blood away from the heart to capillaries
2. **Capillaries:** The fine network of vessels that increases surface area for exchange of nutrients, gases and toxins between blood & interstitial fluid.
3. **Veins:** Return blood from capillaries to the heart

Flow of blood in the Cardiovascular system

- **Step 1:** Blood begins its flow with the right ventricle pumping blood to the lungs. In the lungs, the blood loads O_2 and unloads CO_2

- **Step 2:** Oxygen-rich blood from the lungs enters the heart at the left atrium, then flows down into left ventricle and is then pumped via aorta to the entire body

- **Steps 3:** From the body, blood returns to the heart via superior and inferior vena cava which pour blood into the right atrium of the heart.

The Pulmonary and Systemic Circuits

- Heart is a transport system where the Right side receives oxygen-poor blood from tissues then Pumps blood to lungs to get rid of CO_2, pick up O_2, via **pulmonary circuit**

- **Whereas the** Left side receives oxygenated blood from lungs and pumps blood to body tissues via **systemic circuit**

- Equal volumes of blood are pumped to pulmonary and systemic circuits. However, the left ventricle walls are 3× thicker than right to pumps with greater pressure

Chambers of the Heart

- *Receiving chambers of heart*

1. **Right atrium** - Receives blood returning from systemic circuit
2. **Left atrium** - Receives blood returning from pulmonary circuit

- *Pumping chambers of heart*

3. **Right ventricle** -Pumps blood through pulmonary circuit
4. **Left ventricle** -Pumps blood through systemic circuit

Valves of the Heart

- Two <u>Atrioventricular valves</u> located between atria & ventricles

1. **Tricuspid valve** (right AV valve)
2. **Mitral valve** (left AV valve, bicuspid valve)

- Two <u>**Semilunar (SL) valves**</u> : located between ventricles and major arteries.

3. **Pulmonary semilunar valve:** between right ventricle and pulmonary trunk
4. **Aortic semilunar valve:** between left ventricle and aorta

The **Atrioventricular (AV) valves** separate each atrium and ventricle

- Valves ensure unidirectional blood flow through heart and they open and close only in response to pressure changes

- Two **Atrioventricular valves** located between atria & ventricles prevent backflow of blood into atria when ventricles contract

- The "lub-dup" sound of a heart beat is caused by the recoil of blood against the AV valves (lub) then against the semilunar (dup) valves

---

The **Semilunar (SL) valves** separate each ventricle from major arteries

- The 2 **semilunar valves** control blood flow to the aorta and the pulmonary artery

- Two **Semilunar (SL) valves** : located between ventricles and major arteries. Prevent backflow from major arteries back into ventricles

- Backflow of blood through a defective valve causes a **heart murmur**

- The heart contracts and relaxes in a rhythmic cycle called the cardiac cycle
- The contraction, or pumping, phase is called **systole**. The relaxation, or filling, phase is called **diastole**
- The **heart rate**, also called the pulse, is the number of beats per minute
- The **stroke volume** is the amount of blood pumped in a single contraction
- The **cardiac output** is the volume of blood pumped into the systemic circulation per minute and depends on both the heart rate and stroke volume

---

Pathway of blood through Heart

- **Right side of the heart**
  - Superior & inferior vena cava and coronary sinus
  - Right atrium
  - Tricuspid valve
  - Right ventricle
  - Pulmonary semilunar valve
  - Pulmonary trunk & arteries
  - Lungs

- **Left side of the heart**
  - Four pulmonary veins
  - Left atrium
  - Mitral valve
  - Left ventricle
  - Aortic semilunar valve
  - Aorta
  - Systemic circulation

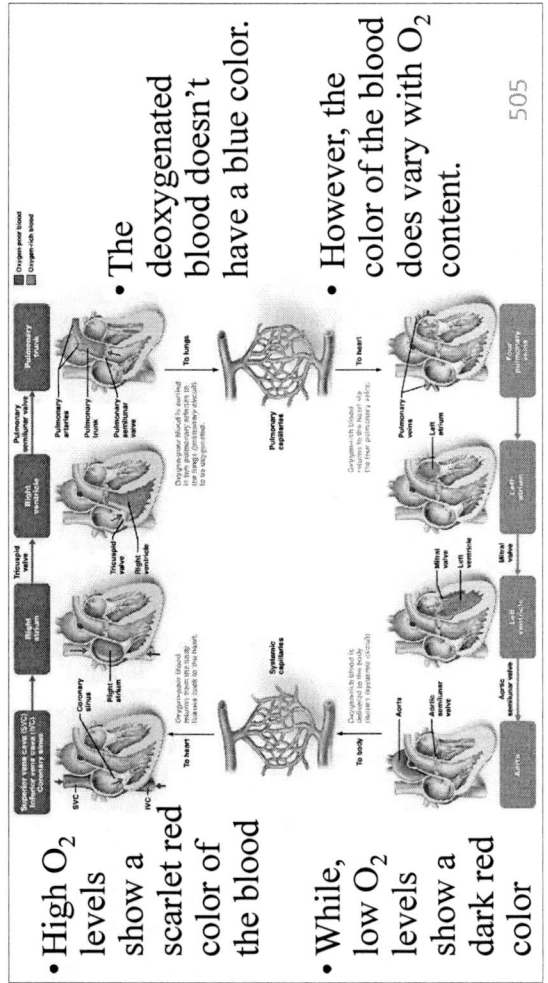

- High O_2 levels show a scarlet red color of the blood
- While, low O_2 levels show a dark red color

Coronary circulation- by the heart, for the heart

- Functional blood supply to heart muscle itself
- Shortest circulation in the body and is delivered when heart is relaxed
- *Coronary arteries* arise from base of aorta and supply arterial blood to heart
- Prolonged coronary blockage causes **Myocardial infarction** (heart attack).
- Areas of cell death are repaired with noncontractile scar tissue

Heart's rhythmic beat- the Sinus rhythm

- Some cardiac muscle cells are self-excitable, meaning they contract without any signal from the nervous system
- The **sinoatrial (SA) node**, or *pacemaker*, sets the rate and timing at which cardiac muscle cells contract
- Impulses from the **SA** node travel to the **AV** node
- At the AV node, the impulses are delayed and then travel to the Purkinje fibers that make the ventricles contract

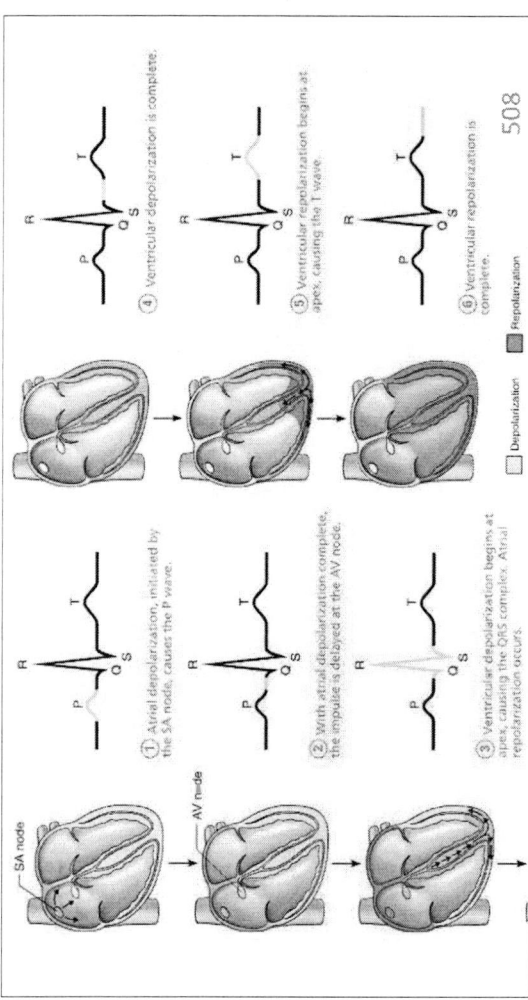

The Sinus Rhythm

- **Electrocardiograph** can detect electrical currents generated by heart

- Impulses that travel during the cardiac cycle can be recorded as an **electrocardiogram (ECG or EKG)**

- Electrodes are placed at various points on body to measure voltage differences

Blood Pressure and changes in it during Cardiac Cycle

- **Blood pressure** is the hydrostatic pressure that blood exerts against the wall of a vessel

- **Systolic pressure** is the pressure in the arteries during ventricular systole; it is the highest pressure in the arteries

- **Diastolic pressure** is the pressure in the arteries during diastole; it is lower than systolic pressure

- A **pulse** is the rhythmic bulging of artery walls with each heartbeat

Regulation of Blood Pressure

- Blood pressure is determined by cardiac output and resistance due to constriction of arterioles

- **Vasoconstriction** is the contraction of smooth muscle in arteriole walls; it increases blood pressure

- **Vasodilation** is the relaxation of smooth muscles in the arterioles; it causes blood pressure to fall

- Vasoconstriction and vasodilation help maintain adequate blood flow as the body's demands change

Measuring Blood pressure

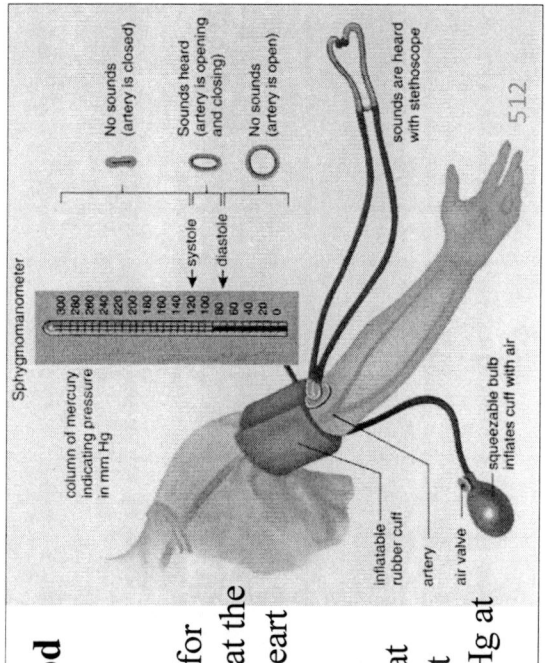

- Blood pressure is generally measured for an artery in the arm at the same height as the heart

- Blood pressure for a healthy 20 year old at rest is 120 mm Hg at systole and 70 mm Hg at diastole

Functions of Blood

- **Blood** is the life-sustaining transport vehicle of the cardiovascular system. Functions include **transport, regulation & protection**

1. **Transport** functions include:
   - Delivering O_2 and nutrients to body cells
   - Transporting metabolic wastes to lungs and kidneys for elimination
   - Transporting hormones from endocrine organs to target organs

2. **Regulation** functions include:
   - Maintaining body temperature by absorbing and distributing heat
   - Maintaining normal pH using buffers; alkaline reserve of bicarbonate ions

3. **Protection** functions include:
   - Preventing blood loss: Plasma proteins and platelets in blood initiate clot formation
   - Preventing infection: Agents of immunity such as Antibodies & white blood cells are carried in blood

Components of Blood: Plasma & Cellular elements

- Solutes found in **blood plasma** are inorganic salts in the form of dissolved ions, sometimes called electrolytes
- Another important class of solutes is the plasma proteins, which influence blood pH, osmotic pressure, and viscosity, lipid transport, immunity, and blood clotting

- **Cellular elements** consists of:
  1. **Red blood cells** (erythrocytes) that transport oxygen
  2. **White blood cells** (leukocytes) that function in defense
  3. **Platelets** (fragments of cells) that are involved in clotting

Red blood cells (erythrocytes)

- Red blood cells, or **erythrocytes**, are by far the most numerous blood cells. They transport oxygen throughout the body
- They contain **hemoglobin**, the iron-containing protein that transports oxygen.
- Globin is composed of 4 polypeptide chains (2alpha, 2 beta)& each heme's central iron atom binds one O_2
- Each Hb molecule can transport four O_2. Each RBC contains 250 million Hb molecules
- **Erythropoietin (EPO)**: hormone stimulates formation of RBCs

White blood cells (leukocytes)

- There are five major types of white blood cells, or **leukocytes**: monocytes, neutrophils, basophils, eosinophils, and lymphocytes

- They function in defense by phagocytizing bacteria and debris or by producing antibodies

- In case of an infection, the body responds by increasing the no. of WBC's. A number over 11,000 per μl of blood, is indicative of an infection & also of the body's response to that infection (inflammation)

Platelets and their role in Hemostasis

- Platelets are fragments of cells and function in blood clotting

- When the endothelium of a blood vessel is damaged, the clotting mechanism (**Hemostasis**) begins

- Three steps involved
  - **Step 1: Vascular spasm**
  - **Step 2: Platelet plug formation**
  - **Step 3: Coagulation (blood clotting)**

Stem Cells and the Replacement of Cellular Elements

- The cellular elements of blood wear out and are replaced constantly throughout a person's life

- Erythrocytes, leukocytes, and platelets all develop from a common source of **stem cells** in the bone marrow

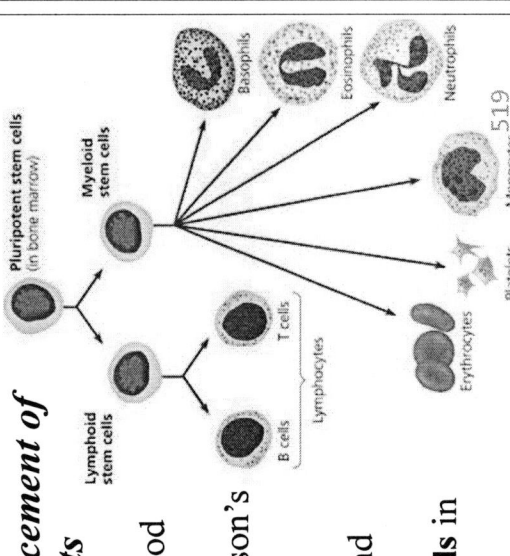

Cardiovascular Disease

- Cardiovascular diseases are disorders of the heart and the blood vessels

- One type of cardiovascular disease, **atherosclerosis**, is caused by the buildup of plaque deposits within arteries

- **A heart attack** is the death of cardiac muscle tissue resulting from blockage of one or more coronary arteries

- **A stroke** is the death of nervous tissue in the brain, usually resulting from rupture or blockage of arteries in the head

Prevention of Cardiovascular Disease

- Cholesterol is a major contributor to atherosclerosis
- **Low-density lipoproteins (LDLs)** are associated with plaque formation; these are "bad cholesterol"
- **High-density lipoproteins (HDLs)** reduce the deposition of cholesterol; these are "good cholesterol"
- **Hypertension**, or high blood pressure, promotes atherosclerosis and increases the risk of heart attack and stroke
- Hypertension can be reduced by dietary changes, exercise, and medication

Gas exchange

- **Gas exchange** supplies oxygen for cellular respiration and disposes of carbon dioxide
- **Partial pressure** is the pressure exerted by a particular gas in a mixture of gases
- A gas diffuses from a region of higher partial pressure to a region of lower partial pressure (diffusion)
- In the lungs and tissues, O_2 and CO_2 diffuse from where their partial pressures are higher to where they are lower

Respiratory surfaces that help in gas exchange

- Air inhaled through the nostrils passes through the **pharynx** into the **larynx, trachea, bronchi, bronchioles,** and **alveoli**, where gas exchange occurs

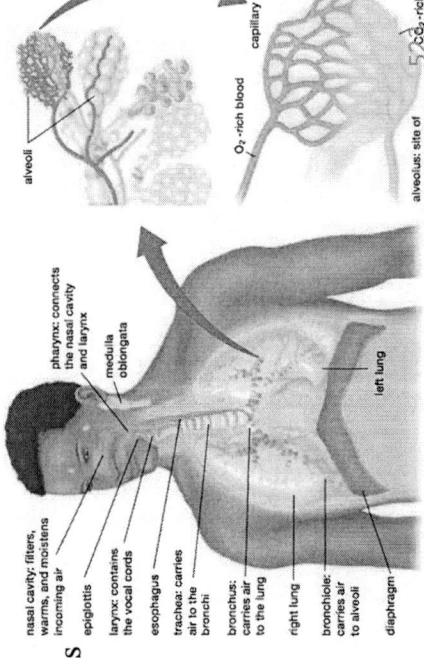

Mammals ventilate their lungs by **negative pressure breathing**, which pulls air into the lungs.

- Lung volume increases as rib muscles & **diaphragm** contract.
- The **tidal volume** is the volume of air inhaled with each breath
- Maximum tidal volume is the **vital capacity**
- After exhalation, a **residual volume** of air remains in the lungs

Breathing control centers

- In humans, the main breathing control centers are in two regions of the brain, the **medulla oblongata** and the **pons**

- The medulla adjusts breathing rate and depth to match metabolic demands and the pons regulates the tempo

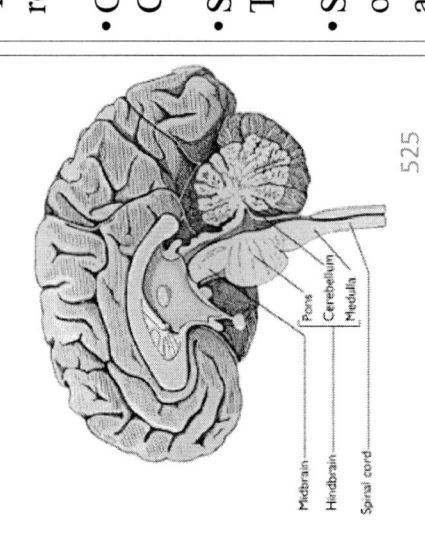

Midbrain
Hindbrain
Spinal cord
Pons
Cerebellum
Medulla

Regulation of Breathing in Humans

- The medulla regulates the rate and depth of breathing in response to pH changes in the cerebrospinal fluid

- Other sensors in the aorta and carotid arteries monitor O_2 and CO_2 concentrations in the blood

- Sensors in the alveoli can detect fluid buildup in the lung tissues. These sensors are thought to trigger rapid, shallow breathing.

- Sensors in your joints & muscles detect movement of your arms or legs, & increase your breathing rate when you're physically active

Respiratory pigments bind and transport gases

- **Respiratory pigments**, proteins that transport oxygen, greatly increase the amount of oxygen that blood can carry

- Arthropods and many molluscs have hemocyanin with copper as the oxygen-binding component

- Most vertebrates and some invertebrates use hemoglobin with iron as the oxygen-binding component

- A single hemoglobin molecule can carry four molecules of O_2

P_{O_2} & Hemoglobin dissociation curve

- The hemoglobin dissociation curve shows that a small change in the partial pressure of oxygen can result in a large change in delivery of O_2

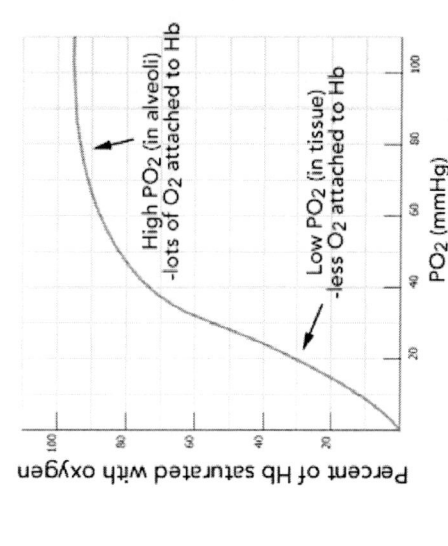

P_{O_2} and hemoglobin dissociation at pH 7.4

pH and hemoglobin dissociation curve

- CO_2 produced during cellular respiration lowers blood pH and decreases the affinity of hemoglobin for O_2; this is called the **Bohr shift**

- Hemoglobin's oxygen binding affinity is inversely related both to acidity and to the concentration of carbon dioxide

pH and hemoglobin dissociation

Bohr shift also known as Bohr effect

- As cells metabolize glucose, they use O_2, causing:

  Oxygen-hemoglobin Dissociation: Exercise

- Increases in P_{CO_2} and H^+ in capillary blood

- Declining blood pH (acidosis) and increasing P_{CO2} cause Hb-O_2 bond to weaken and is referred to as **Bohr effect** or Bohr shift.

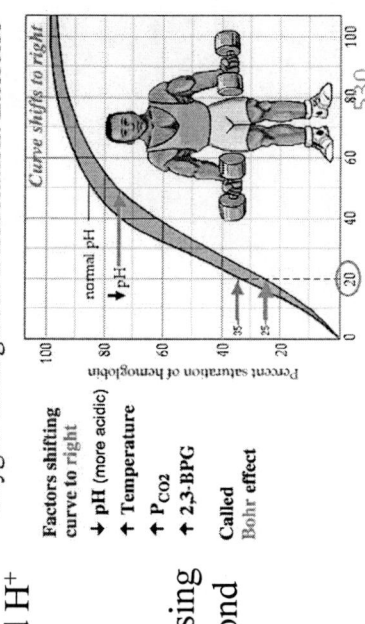

Factors shifting curve to right
↓ pH (more acidic)
↑ Temperature
↑ P_{CO2}
↑ 2,3-BPG

Called Bohr effect

18. Immune system

Immune system

- The **immune system** recognizes foreign bodies and responds with the production of immune cells and proteins
- **Immune system** provides resistance to disease & is made up of two fundamental systems

1. **Innate (nonspecific) defense system:** Constitutes 1st and 2nd lines of defense. *(found in both invertebrates and vertebrates)*
2. **Adaptive (specific) defense system:** 3rd line of defense, attacks particular foreign substances. *(only in vertebrates)*

Innate defense system

1. Innate defense system: *1st & 2nd lines of defense*

- **Innate immunity** is present before any exposure to pathogens and is effective from the time of birth
- *1st line of defense*: external body membranes (skin & mucosae)
- *2nd line of defense*: antimicrobial proteins, phagocytes, inflammation, Natural killer cells, fever
- It is a rapid, nonspecific responses to pathogens.

First Line of Defense

- Skin and mucous membranes produce protective chemicals that inhibit or destroy microorganisms
- **Acid:** Low pH forms an acid mantle & inhibits microbial growth
- **Enzymes:** lysozyme of saliva & enzymes in stomach kill many microbes
- **Mucus:** lines digestive & respiratory tracts, traps microbes
- **Defensins:** antimicrobial peptides that inhibit microbial growth

Second Line of Defense

- Innate system internal defenses

a) Phagocytes
b) Natural killer (NK) cells
c) Inflammatory response (macrophages, mast cells, WBCs, and inflammatory chemicals)
d) Antimicrobial proteins (interferons and complement proteins)
e) Fever

a) Phagocytosis

- White blood cells (leukocytes) engulf pathogens in the body
- A white blood cell engulfs a microbe, then fuses with a lysozyme to destroy the microbe.
- Some examples of Leukocytes that carry phagocytosis are:
    - **Neutrophils**
    - **Eosinophils**
    - **Dendritic cells**
    - **Macrophages** (produced by differentiation of Monocytes)

b) Natural killer (NK) cells

- All cells in the body (except red blood cells) have a class 1 MHC (major histocompatibility complex) protein on their surface
- Cancerous or infected cells no longer express this protein
- **Natural killer (NK) cells** attack these damaged cells by inducing **apoptosis** in cancer cells and virus-infected cells

c) Inflammatory Responses

- Inflammation is body's systemic response to an infection to protect us from foreign organisms, such as bacteria & viruses.
- **Inflammation** is triggered whenever body tissues are injured due to trauma, heat, chemicals, or infections by microorganisms
- Following an injury, **mast cells** release **histamine**, which causes vasodilation. (dilated vessels causes increase in blood supply)
- This allows more phagocytes to enter the tissues.

- **Stages of inflammation**

  1. Inflammatory chemical release (Histamine)
  2. Vasodilation and increased vascular permeability
  3. Phagocyte mobilization
  4. Phagocytes destroy pathogens

- **Benefits of inflammation:**
  - Prevents spread of damaging agents
  - Disposes of cell debris and pathogens
  - Alerts adaptive immune system
  - Sets the stage for repair
- **Four cardinal signs** of acute inflammation:
  1. Redness 2. Heat 3. Swelling 4. Pain
- *Pus*, a fluid rich in white blood cells, dead microbes, and cell debris, accumulates at the site of inflammation

d) Antimicrobial Proteins

- **Antimicrobial proteins** enhance innate defense by attacking microbes directly or by hindering their ability to reproduce
- **Complement and IFN** are 2 important antimicrobial proteins.

1. **Complement system:** consists of ~20 blood proteins that circulate in blood in inactive form
- Provides major mechanism for destroying foreign substances
- Activation enhances inflammation and also directly destroys bacteria

2. **Interferons (IFN):** Cells infected with viruses can secrete IFNs that "warn" healthy neighboring cells.
- IFNs enter neighboring cells, stimulating production of proteins that block viral reproduction and degrade viral RNA.
- IFNs activate NK cells and macrophages, so they indirectly fight cancer
- Artificial IFNs are used to treat disorders such as hepatitis C, genital warts, and multiple sclerosis

e) Fever

- High body temperature is body's systemic response to invading microorganisms

- Leukocytes and macrophages exposed to foreign substances secrete **pyrogens.**

- Pyrogens act on body's thermostat in hypothalamus, raising body temperature. The benefits of having a moderate fever are:
  - It causes liver & spleen to seize iron and zinc (needed by m/o)
  - Increases metabolic rate, which increases rate of repair

Adaptive defense system

2. Adaptive (acquired) defense system: *3rd line of defense* attacks *specific* substances (takes longer to react than innate)

Adaptive immunity, develops after exposure to agents such as microbes & toxins. The 3 branches of adaptive immunity are:

a) **Humoral immunity: B cells**
b) **Cellular immunity: T cells**
c) **Immunity aided by Antigen Presenting cell: APC**

- Adaptive immunity is a slow, specific response to pathogens.

Third Line of Defense

- **Adaptive immune system** eliminates almost any pathogen or abnormal cell in body

- Four characteristics of adaptive immunity:
  - It is **specific**: recognizes and targets specific antigens
  - It has **memory**: mounts a stronger attack to "known" antigens
  - **Amplifies** inflammatory response
  - **Activates** complement system

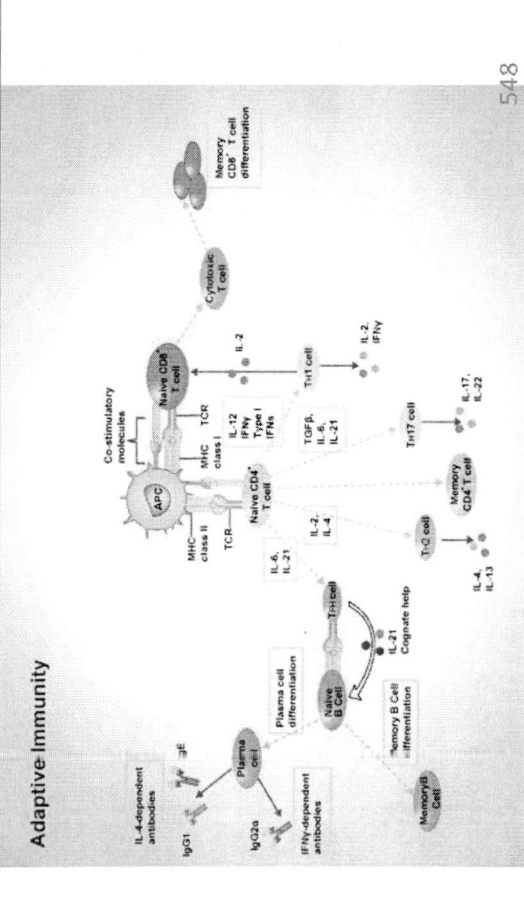

Adaptive Immunity

Lymphocytes - Development, maturation, & activation

- Both **T** and **B** lymphocytes share common development and steps in their life cycles. Five general steps:

1. **Origin**: both lymphocytes originate in red bone marrow
2. **Maturation** - Lymphocytes mature in **primary lymphoid organs** – bone marrow (for B cells) and thymus (for T cells)
3. **Secondary lymphoid organs** - Exported from primary to secondary lymphoid organs (lymph nodes, spleen, etc.)

4. **Antigen encounter & activation:** Naive lymphocyte's first encounter with antigen, triggers lymphocyte to develop further

- Lymphocyte is selected to differentiate into active cell by binding to its specific antigen (referred to as **clonal selection**)

5. **Proliferation and differentiation:** Once selected and activated, lymphocyte proliferates & forms army of exact copies of itself. (referred to as **clones**)

- Most clones become **effector cells** & fight infections while few remain as **memory cells** (for robust secondary immune response)

Three main branches of adaptive system

a) Humoral immunity: In HIR, B cells freely circulate in body fluids.

- When they come across an Antigen, they bind and inactivate them and **mark them** for destruction by phagocytes.

- In HIR, B cells undergo differentiation & give rise to **plasma cells**.

- **Plasma cells** secrete proteins called **immunoglobulins or antibodies** (ex: IgM, IgA, IgD, IgG, and IgE)

b) Cellular Immunity: T cells provide defense against intracellular antigens. Ex: cells infected with viruses or bacteria, or cancerous & abnormal cells.

- Some T cells directly kill cells; others release chemicals that regulate immune response. 4 kinds of T cells :

1. **Helper T cells:** activate B cells, other T cells, & macrophages
2. **Regulatory T cells**, suppresses immune response
3. **Memory T cells**, provide immunological memory
4. **Cytotoxic T cells** (T_C) destroys cells that have been infected

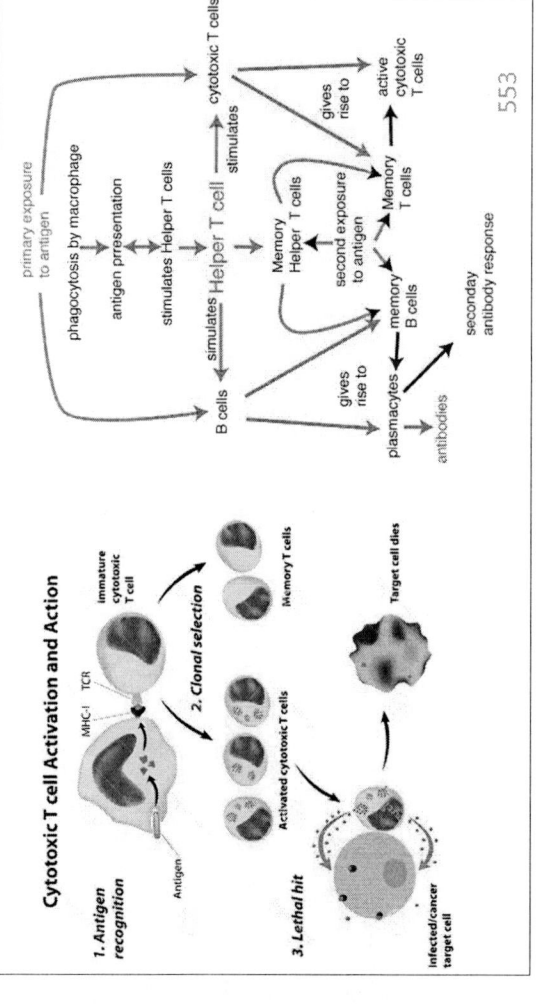

c) **Antigen-presenting cells (APCs):** Engulf antigens & presents fragments of antigens to T cells for recognition. 3 Major types

1. **Dendritic cells**- Phagocytize pathogens that enter tissues, then enter lymphatics to present antigens to T cells in lymph node

2. **Macrophages**- Present antigens to T cells, activates T cells and macrophages. *Activated macrophages become voracious phagocytic killers*

3. **B lymphocytes** - Present antigens to specialized helper T cell to assist their own activation

Antigen presentation

- Antigen presentation is vital for activation of naive T cells and normal functioning of effector T cells

- T cells respond to processed fragments of antigens displayed on surfaces of cells by MHC proteins.

- 2 classes of MHC (major histocompatibility complex)proteins

 – **Class I MHC proteins:** displayed by all cells except RBCs
 – **Class II MHC proteins:** displayed by APCs

Antigens and Antibodies

- **Antigens**: substances that can mobilize adaptive defenses and provoke an immune response

- Most are large, complex molecules not normally found in body

- Antigen-antibody complexes do not destroy antigens; they prepare them for destruction by innate defenses

- **Antibodies** go after extracellular pathogens; they do not invade solid tissue unless lesion is present

Primary immune response:

- Cell proliferation and differentiation upon exposure to antigen for the first time

- During this time, effector B cells called **plasma cells** are generated, and T cells are activated to their effector forms

- Lag period: 3 to 6 days
- Peak levels of plasma antibody are reached in 10 days
- Antibody levels then decline

Secondary immune response: Immunological Memory

- **Secondary immune response:** Re-exposure to same antigen gives faster, more prolonged, more effective response

- Memory cells provide **immunological memory**
- Respond within hours, not days
- Antibody levels peak in 2 to 3 days at much higher levels
- Antibodies bind with greater affinity
- Antibody level can remain high for weeks to months

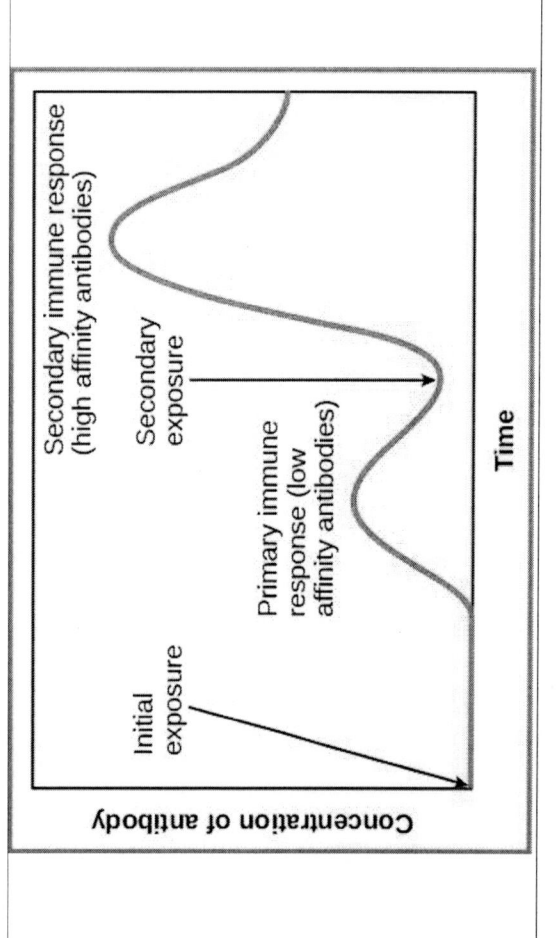

Adaptive Immunity

Natural
Vs
Artificial

Active
Vs
Passive

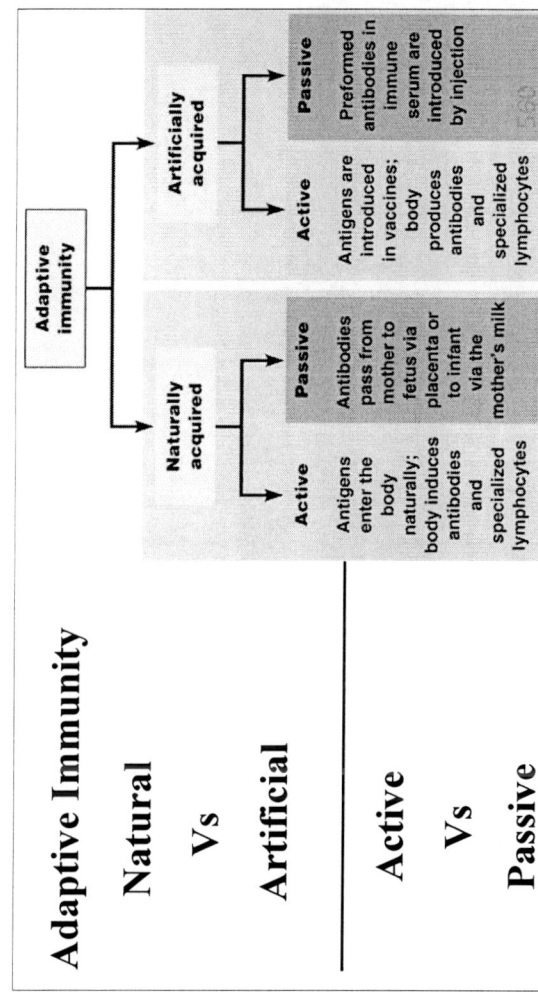

Active and Passive Humoral Immunity

- **Active humoral immunity** occurs when B cells encounter antigens & produce specific antibodies against them. Two types:

1. Naturally acquired: formed in response to an actual bacterial or viral infection

2. Artificially acquired: formed in response to a **vaccine** of dead or attenuated pathogens.

- Provides antigenic determinants that are immunogenic and reactive & spares us symptoms of primary response

- **Passive humoral immunity** occurs when ready-made antibodies are introduced into body
  - B cells are not challenged by antigens
  - Immunological memory does not occur
  - Protection ends when antibodies degrade

- **Two types** of **passive** humoral immunity

1. Naturally acquired: antibodies delivered to fetus via placenta(IgG) or to infant through milk(IgA)

2. Artificially acquired: injection of serum, e.g.: gamma globulin

Immune Rejection

- Cells transferred from a donor to a recipient can elicit an immune system response in the recipient to attack it's own self.

- Antigens on red blood cells determine whether a person has blood type A (A antigen), B (B antigen), AB (both A and B antigens), or O (neither antigen)

- Antibodies to nonself blood types exist in the body

- Transfusion with incompatible blood leads to destruction of the transfused cells.

- Recipient-donor combinations can be fatal or safe

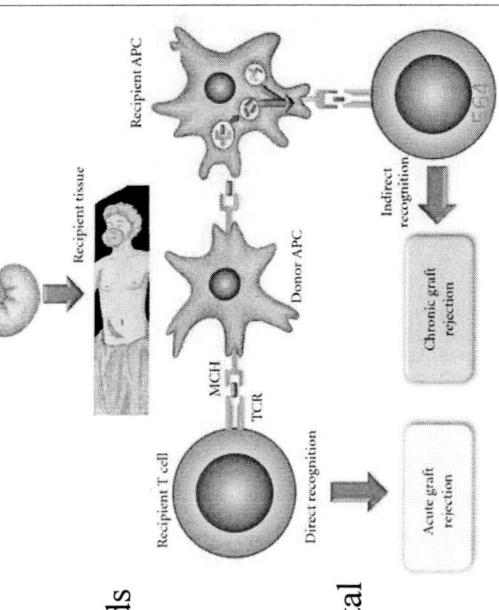

Disruption of immune system

- Disruption in immune system function can have minor to fatal consequences.
- In some cases it can elicit a disease, in others, worsen an already existing one.
- The 3 kinds of disruption in Immune system that effect us most are Immunodeficiency, Autoimmune diseases and Allergens

Immunodeficiency's

- **Immunodeficiency**: congenital or acquired conditions that impair function or production of immune cells or molecules
- **Congenital- Severe combined immunodeficiency (SCID)** syndrome: genetic defect with marked deficit in B & T cells
- Defective **adenosine deaminase enzyme** allows accumulation of metabolites lethal to T cells; fatal if untreated

2. (a) Acquired- Hodgkin's disease is an acquired immunodeficiency that causes cancer of B cells, which depresses lymph node cells and leads to immunodeficiency

Hodgkin's disease (a lymphoma) spreads in a stepwise fashion from one lymph node to another and later spreads through blood

- Painless swelling of the lymph nodes in the neck, armpits, or groin

2. (b) Acquired immune deficiency syndrome (AIDS) by HIV

- **Human immunodeficiency virus (HIV)** cripples immune system by interfering with activity of helper T cells
- Characterized by severe weight loss, night sweats, and swollen lymph nodes
- Opportunistic infections occur, including *Pneumocystis* pneumonia and Kaposi's sarcoma
- HIV destroys T_H cells, thereby depressing cellular immunity

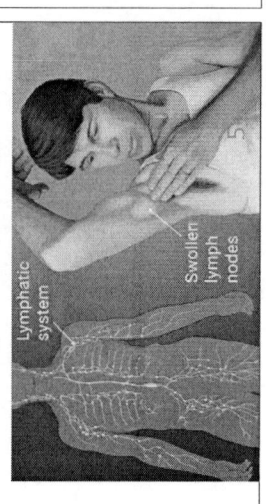

Human immunodeficiency virus

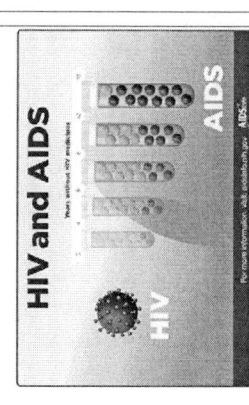
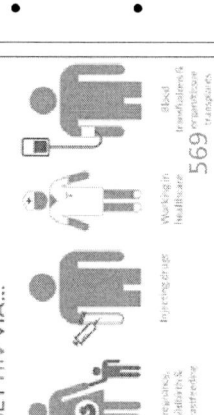

- HIV is transmitted via body fluids: blood, semen, and vaginal secretions
- HIV can enter the body via: Blood transfusions; blood-contaminated needles; sexual intercourse and oral sex; also from mother to fetus
- HIV multiplies in lymph nodes throughout asymptomatic period, ~10 years if untreated
- Symptoms begin when immune system collapses. Virus also can invade brain, leading to dementia

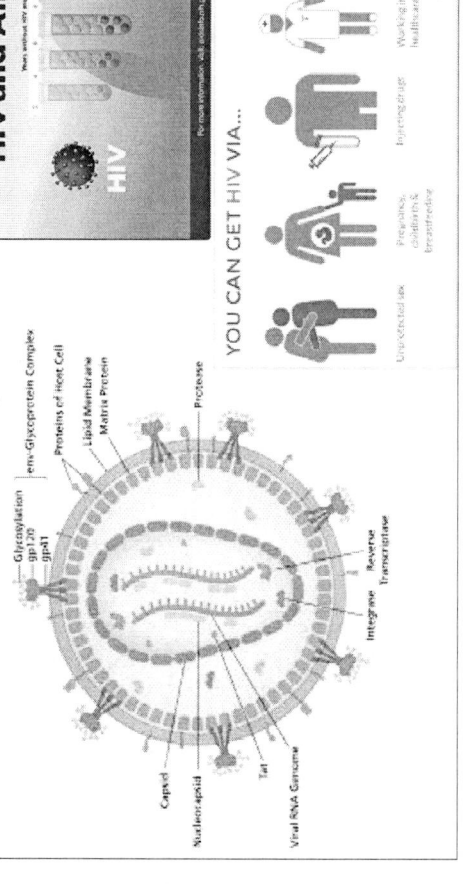

- **HIV** enters cells and uses *reverse transcriptase* to produce DNA from its viral RNA.
- HIV reverse transcriptase is prone to frequent errors, leading to high mutation rate and resistance to drugs.
- The DNA copy (a **provirus**) directs host cell to make viral RNA and proteins, enabling virus to reproduce
- No cure for AIDS found; four major classes of antivirals in combination help but can fail as virus becomes resistant

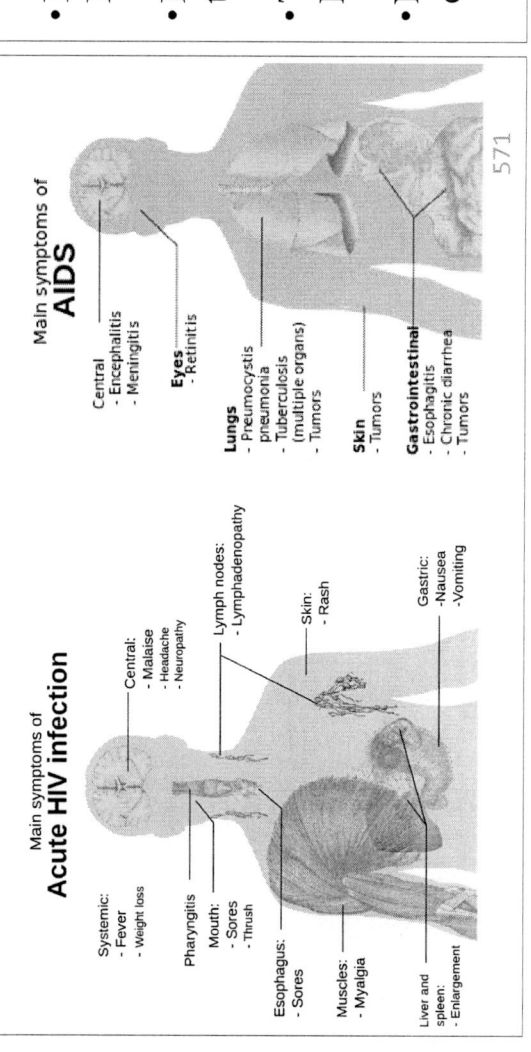

Autoimmune Diseases

- **Autoimmune disease** results when immune system loses ability to distinguish self from foreign
- **Autoimmunity**: The immune response of producing autoantibodies & T_C cells that destroys body tissues. Examples:
  - Rheumatoid arthritis: destroys joints
  - Myasthenia gravis: impairs nerve-muscle connections
  - Multiple sclerosis: destroys white matter myelin
  - Type 1 diabetes mellitus: destroys pancreatic cells

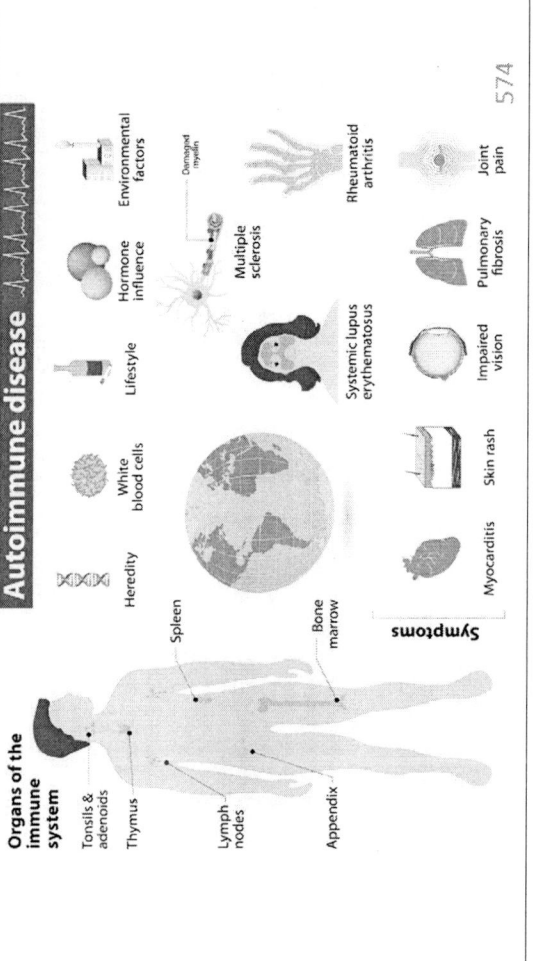

- **Cause of Autoimmune diseases** - Failure of self-tolerance
- Autoimmunity may be caused by self-reactive lymphocytes that are activated by:
  - Foreign antigens that resemble self-antigens
  - Appearance of new self-antigens, generated by gene mutations
- **Treatment of autoimmune diseases**
  - Suppress entire immune system- Anti-inflammatory drugs, such as corticosteroids

Allergy's

- Allergies' are Immune system's hypersensitive response to harmless threat. Symptoms include runny nose, watery eyes, sneezing, itchy skin, hives, coughing, asthma.
- Some examples of this exaggerated responses to allergens are
  - Contact dermatitis (e.g., poison ivy)
  - Injected allergens (example: bee sting)
  - Mismatched blood transfusion reaction
  - Food allergens (peanut)

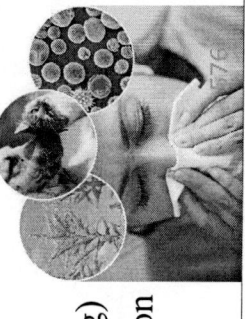

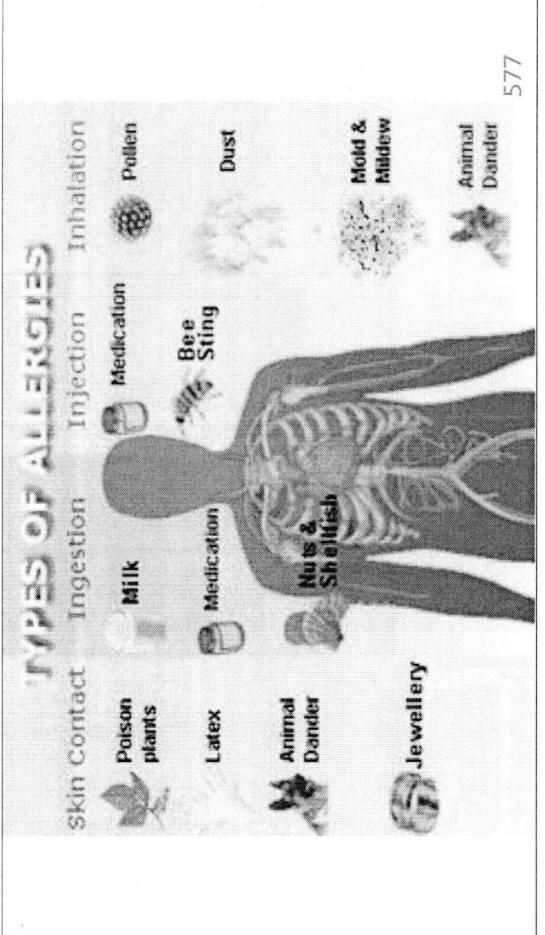

Signs and symptoms of Anaphylaxis

- Swelling of the conjunctiva
- Runny nose
- Central nervous system
  - lightheadedness
  - loss of consciousness
  - confusion
  - headache
  - anxiety
- Respiratory
  - shortness of breath
  - wheezes or stridor
  - hoarseness
  - pain with swallowing
  - cough
- Swelling of lips, tongue and/or throat
- Heart and vasculature
  - fast or slow heart rate
  - low blood pressure
- Skin
  - hives
  - itchiness
  - flushing
- Gastrointestinal
  - crampy abdominal pain
  - diarrhea
  - vomiting
- Pelvic pain
- Loss of bladder control

- **Anaphylaxis**: An extreme allergic reaction to an antigen (e.g., a bee sting) to which the body has become hypersensitive.

- It may cause a range of symptoms.

- Treatment: epinephrine

Immune System Evasion by Pathogens

a) Attack on the Immune System

- Human immunodeficiency virus (HIV) infects helper T cells

- The loss of helper T cells impairs both the humoral and cell-mediated immune responses and leads to AIDS

- HIV eludes the immune system because of antigenic variation and an ability to remain latent while integrated into host DNA

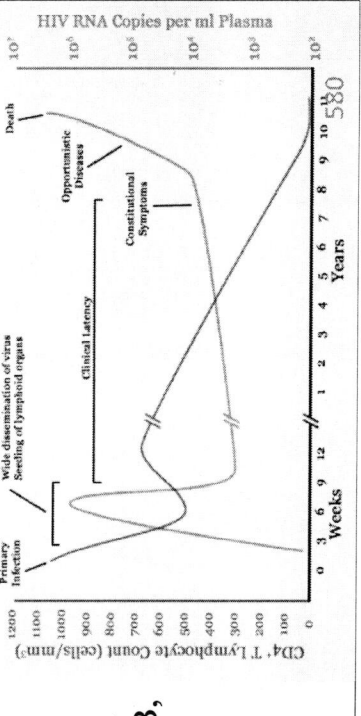

b) *Latency*

- Some viruses may remain in a host in a dormant (inactive state) called latency.

- E.g.: Herpes simplex viruses, HIV, Hepatitis B, Rubella

c) Antigenic Variation

- Through antigenic variation, some pathogens are able to change epitope expression and prevent recognition

- The human influenza virus mutates rapidly, and new flu vaccines must be made each year

19. Study of Ecology

Ecology - the study of interactions

- **Ecology** is the study of the interactions between the organisms and also between the organisms and their environment
- These interactions influence the dispersal and eventually the distribution of organisms
- Ecology can be studied at levels ranging from an individual to a global level
- Ecology provides the scientific understanding that underlies environmental issues

Ecology at 6 levels

1. Organismal ecology studies how an organism's Anatomy, physiology, & behavior is affected in response to environmental challenges.

- At this level we also study the adaptations that enable individuals to live in specific habitats

2. Population ecology: Population is defined a group of individuals (of same species) living in a certain geographic area. **Population ecology** studies factors that affect individuals of a species in a certain area

3. Community ecology: Community is defined as an assembly of several different populations (of different species) living in a certain geographic area.

Community ecology deals with the interaction between different species in a community

4. Ecosystem ecology: Studies the flow of energy & recycling of chemical nutrients among plants, animals & microorganisms.

At this level interactions between Biotic & Abiotic factors (H20, chemicals, plants, animals, microbes) are most widely studied.

5. Landscape ecology
deals with improving the ecological processes in a certain ecosystem in order to make it most efficient.

Example of some ecological processes are: water cycles, energy flow & biogeochemical cycles.

6. Global ecology:
is the study of the interactions among the Earth's ecosystems, land, atmosphere and oceans.

It examines the influence of energy and materials on organisms across the biosphere

Geographic distribution of species

- It has been found that Interactions between organisms and the environment limits the distribution of species

- The two kinds of factors that determine distribution: **biotic**, or living factors, and **abiotic**, or nonliving factors

- Among biotic and abiotic factors, lie several multiple factors that determine the distribution of a species.

Biotic and Abiotic factors

- Below are a few examples of Biotic and Abiotic factors that influence distribution of a species

1. Dispersal
2. Predation
3. Behavior
4. Climate
5. Competition
6. Water & Oxygen
7. Sunlight & temperature
8. Chemical nutrients

Community interactions may help, harm, or have no effect on the species

- A **community** is an assembly of populations of different species living in a certain geographic area

- The relationships between species in a community are called **interspecific interactions**

- Some examples of such interactions are competition, predation, herbivory, and symbiosis (parasitism, mutualism, and commensalism)

Interspecific interactions

- Interspecific interactions affect the survival and reproduction of species, and such effects can be summarized as

  positive (+)   negative (−)   no effect (0)

Competition

- **Interspecific competition** (−/− interaction) occurs when species compete for a resource in short supply

- Strong competition can lead to **competitive exclusion**, local elimination of a competing species

- A species total use of biotic and abiotic resources is called the species' **ecological niche** and can also be thought of as an organism's ecological role

Competitive exclusion or Resources partitioning

- For ecologically similar species that would otherwise face the threat of competitive exclusion, they can learn to coexist in a community but only if there were a few significant differences in their niches

- In other words if they had a way of partitioning the resources they would survive together.

- **Resources partitioning** is differentiation of ecological niches, enabling similar species to coexist in a community.

Resources partitioning

- E.g.: a Tiger & a Lion, or a Lion and a Hyena which feed on the same preys, learn to coexist, sometimes by marking their territories and confining to those regions and other times by sharing their kill.

- Such behavior makes sure they don't drive each other into extinction.

Predation

- **Predation** (+/− interaction) refers to interaction where one species, the predator, kills and the other, the prey gets eaten.

- Such an interaction is positive for the predator but negative for the prey.

- The adaptations that give predators an upper hand in such an interaction are claws, teeth, fangs, and poison

Claws, Teeth, Fangs vs Hiding, Herds, sounding Alarm

- In order to avoid the preys from being completely wiped out by predators the natural selection has endowed preys with certain defensive adaptations.

- Behavioral defenses include hiding, forming herds, danger alarm and some morphological defense adaptations

- **Cryptic coloration**, or camouflage, makes prey difficult to spot

Herbivory

- **Herbivory** (+/− interaction) refers to an interaction in which an herbivore eats parts of a plant or alga

- Such an interaction is positive for the Herbivore but negative for the Plant.

- Herbivory has led to evolution of plants mechanical & chemical defenses such as: thorns, poisons chemical signaling,, Idioblasts, housing aggressive ants

Plant's Counter defense

1. Thorns: pointy branches or stems that stab the hebivore
2. Chemical signaling (volatile organic compounds) to warn other plants
3. Housing and feeding aggressive ants that protect plants against Herbivores
4. Plants poisons (like castor bean-derived ricin)
5. Idioblasts (cells that act as landmines): they fire prickly calcium oxalate crystals into the mouths of predators and then release an enzyme that is very similar to reptilian venom

Symbiosis – of 3 types- Mutualism, Commensalism, parasitism

- **Symbiosis** is a relationship where two or more species live in direct & intimate contact with one another

1. Mutualism or Mutualistic symbiosis, (+/+ interaction), is an interspecific interaction that benefits both species

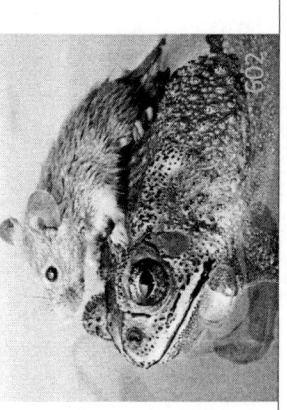

- A mutualism can be such where one species cannot survive without the other or where both species can survive alone

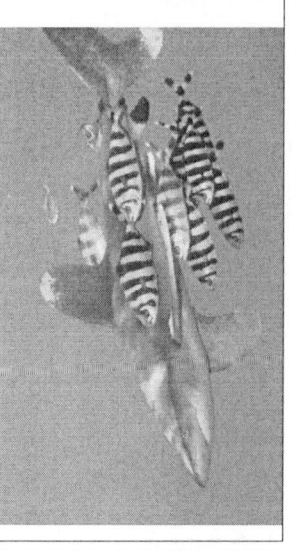

2. Commensalism (+/0 interaction), is when one species benefits and the other is apparently unaffected

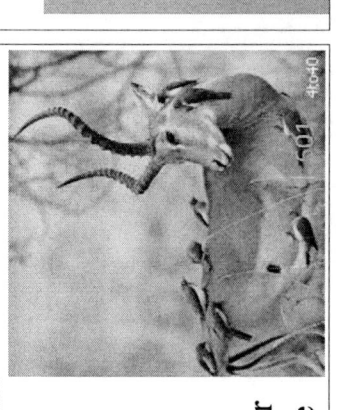

- Commensal interactions are hard to document in nature because any close association likely affects both species

3. Parasitism (+/– interaction), is when one organism, the **parasite**, derives nourishment from another organism, its **host**, which is harmed in the process

- Parasites that live within the body of their host are called **endoparasites**.

- Parasites that live on the external surface of a host are called **ectoparasites**

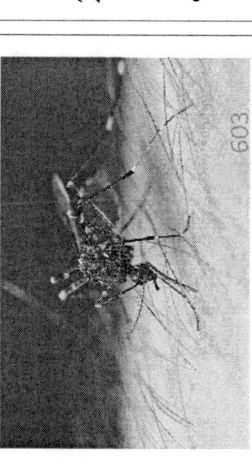

What determines community structure ?

- Two fundamental features that determine a community's structure are species diversity and trophic structure

1. Species diversity of a community is the variety of organisms that make up the community

2. Trophic structure is the feeding relationships between organisms in a community

- The trophic structure in a community are linked with the help of a food chain.

Food web

- **A food web** a complex assembly of interlocking and interdependent food chains.

- **Food webs show** complex trophic interactions

- Species may play a role at more than one trophic level.

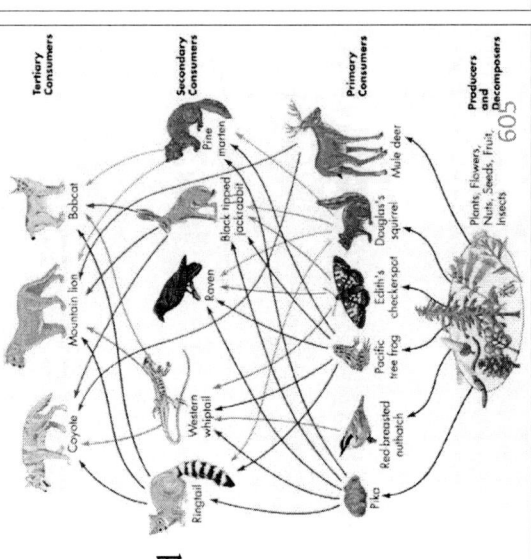

Length of food chains

- The length of food chains are limited and the 2 hypotheses that attempt to explain this limitation are: the energetic hypothesis and the dynamic stability hypothesis

- The **energetic hypothesis** suggests that length is limited by inefficient energy transfer.

- The **dynamic stability hypothesis** proposes that long food chains are less stable than short ones.

Some species exert a strong control on a community's structure

- Certain species have a huge impact on a community's structure

- The 4 kinds of species that most influence a community structure are Dominant, Invasive, Keystone & Foundation

- Such species are either highly abundant or play a critical role in the community dynamics

- Dominant and keystone species exert the strongest controls on community structure

Dominant species

1. **Dominant species** are those that are most abundant or have the highest biomass

- Biomass is the total mass of all individuals in a population

- Dominant species exert powerful control over the occurrence and distribution of other species.

- Either by being most competitive in exploiting resources or by being most successful at avoiding predators

Invasive species

2. Invasive species: can be any kind of living organism—a plant, insect, fish, fungus, bacteria or animal, that is not native to an ecosystem and which causes harm.

- The direct threats of invasive species are:
  - Preying on native species
  - Depleting resources of native species
  - Causing disease
  - Preventing native species from reproducing

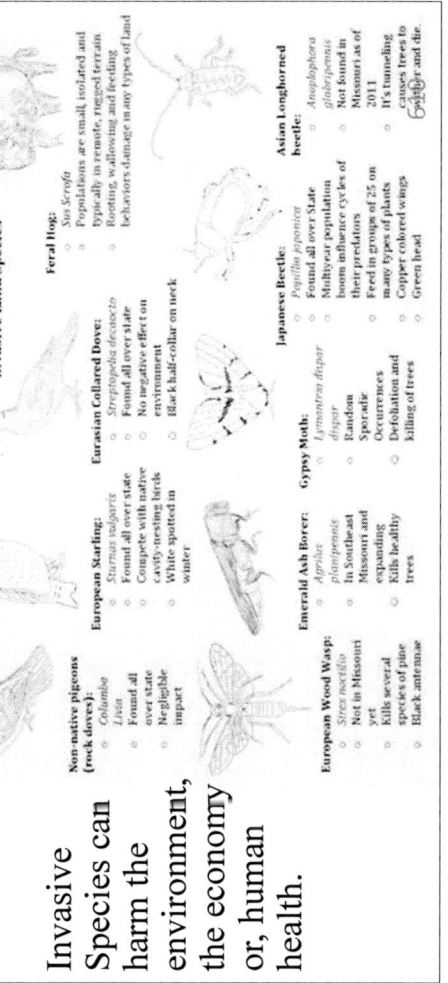

WANTED
Invasive Land Species

Non-native pigeons (rock doves):
- *Columba livia*
- Found all over state
- Negligible impact

European Starling:
- *Sturnus vulgaris*
- Found all over state
- Compete with native cavity-nesting birds
- White spotted in winter

Eurasian Collared Dove:
- *Streptopelia decaocto*
- Found all over state
- No negative effect on environment
- Black half-collar on neck

Feral Hog:
- *Sus Scrofa*
- Populations are small, isolated and typically in remote, rugged terrain
- Rooting, wallowing and feeding behaviors damage many types of land

European Wood Wasp:
- *Sirex noctilio*
- Not in Missouri yet
- Kills several species of pine
- Black antennae

Emerald Ash Borer:
- *Agrilus planipennis*
- In Southeast Missouri and expanding
- Kills healthy trees

Gypsy Moth:
- *Lymantria dispar dispar*
- Random Sporadic Occurrences
- Defoliation and killing of trees

Japanese Beetle:
- *Popillia japonica*
- Found all over State
- Multiyear population boom influence cycles of their predators
- Feed in groups of 25 on many types of plants
- Copper colored wings
- Green head

Asian Longhorned beetle:
- *Anoplophora glabripennis*
- Not found in Missouri as of 2011
- It's tunneling causes trees to wither and die

Invasive Species can harm the environment, the economy or, human health.

Keystone species

3. Keystone species exert strong control on a community by their ecological roles, or niches

- For e.g.: Great white sharks scavenge the sea floor to feed on dead carcasses. In doing so, they help prevent the spread of disease that could be devastating.

- In contrast to dominant species, they are not necessarily abundant in a community

Foundation species

4. Foundation species (ecosystem "engineers") cause physical changes in the environment and strongly affect community structure. For e.g.: Beavers, by making beaver dams.

- Some foundation species act as **facilitators** that have positive effects on survival and reproduction of some other species in the community

Bottom-Up and Top-Down Controls

- The **bottom-up model** of community organization proposes a unidirectional influence from lower to higher trophic levels.

- In this case, presence or absence of mineral nutrients determines community structure, including abundance of primary producers

- The **top-down model** proposes that control comes from the trophic level above. In this case, predators control herbivores, which in turn control primary producers

Disturbance influences species diversity & composition

- For a long time, ecologists favored the view that communities are in a state of equilibrium

- However, recent evidence has suggested that communities live in a **nonequilibrium model.**

- According to a nonequilibrium model, communities constantly change after having being affected by **disturbances**

- A disturbance is an event that changes a community, removes organisms from it, and alters resource availability

Levels of Disturbance's

- Fire & floods are the most significant disturbances in a terrestrial ecosystem.

- The **intermediate disturbance hypothesis** suggests that **moderate** levels of disturbance can foster greater diversity than either high or low levels of disturbance

- **High** levels of disturbance exclude many slow-growing species whereas, **low** levels of disturbance allow dominant species to exclude less competitive species

Ecological Succession

- Ecological succession is the gradual process by which ecosystems change and develop over time.

- Two main types of succession are primary and secondary.

- **Primary succession** is the series of community changes which occur on an entirely new habitat which has never been colonized before.

- For example, a newly dig out rock face or sand dunes. **Primary succession** occurs where no soil exists when succession begins

Facilitate, Inhibit or Tolerate

- Early-arriving species and later-arriving species may be linked in one of three processes:
  - Early arrivals may facilitate appearance of later species by making the environment favorable
  - They may inhibit establishment of later species
  - They may tolerate later species but have no impact on their establishment

- **Secondary succession** is the series of community changes which take place on a previously colonized, but disturbed or damaged habitat.

- For example, the forest populating again after a fire.

- A secondary succession begins in an area where soil remains after a disturbance

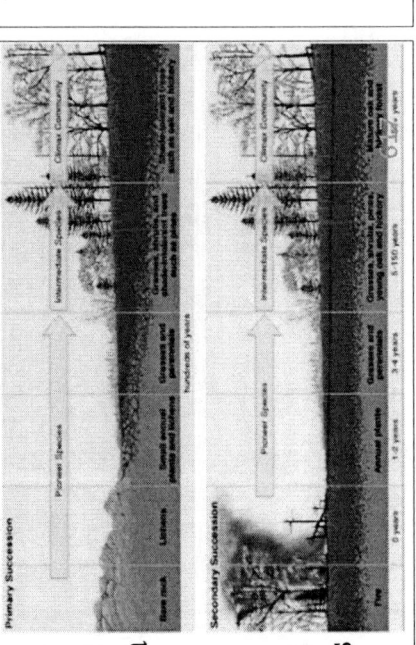

Biogeographic factors affect community biodiversity

- **Latitude and area** are two key factors that affect a community's species diversity (biodiversity).

1. **Climate** is likely the primary cause of the biodiversity.

- Two main climatic factors correlated with biodiversity are solar energy and water availability which can be measured by the rate of evapotranspiration

- **Evapotranspiration** is evaporation of water from soil plus transpiration of water from plants

2. The **species-area curve** quantifies the idea that, all other factors being equal, a larger geographic area has more species

- Studies of species richness on the Galápagos Islands support the prediction that species richness increases with island size

Understanding pathogen life cycles and human disease in a community ecology

- Ecological communities are affected by **pathogens** (disease-causing microbes) which exist as viruses, viroid's, and prions.

- Pathogens can very quickly & extensively wipe out an entire community, as happened during flu epidemic of 1918, that killed approx. 30 million people (more than world war I)

- Human activities are transporting pathogens around the world at unprecedented rates

Community Ecology and Zoonotic Pathogens

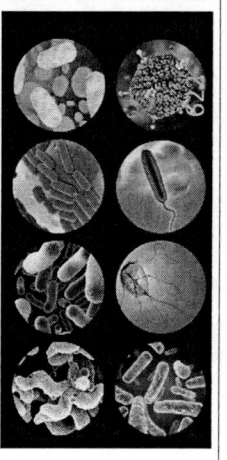

- **Zoonoses**: Infectious diseases of animals that are transferred to humans via **Zoonotic** pathogens.

- The transfer of **pathogens** can be direct or through an intermediate species called a vector

- Many of today's emerging human diseases such as Ebola Virus, Influenza & Salmonellosis are zoonoses

Zoonoses

- Zoonoses can be caused by a range of disease pathogens such as bacteria, fungi, parasites plasmodium, viruses & prions

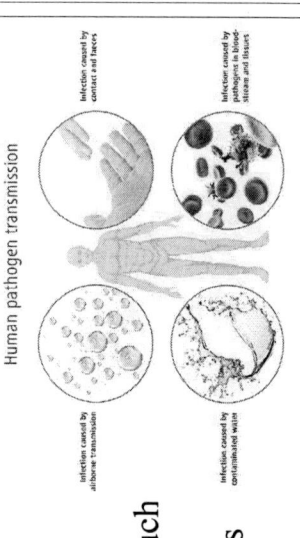

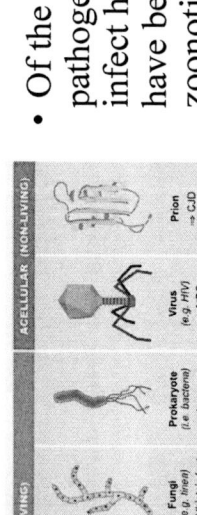

- Of the 1,415 pathogens known to infect humans, 61% have been found to be zoonotic.

20. Ecosystem Ecology

Ecosystems

- An ecosystem studies interactions that living beings have with one another and also with the non-living components in a certain area.

- The two main processes of an ecosystem are: *energy flow* and *chemical cycling*

- Energy flows through ecosystems while nutrients/ chemicals are recycled

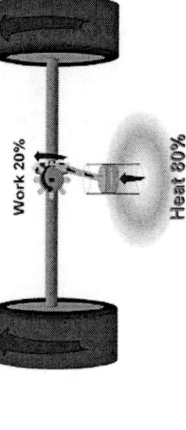

Laws that govern Energy flow & Chemical cycling

- *Conservation of Energy*

- The 1st law of thermodynamics states that energy cannot be created or destroyed, only transformed.

- Energy enters an ecosystem as solar radiation, is conserved, and is lost from organisms as heat

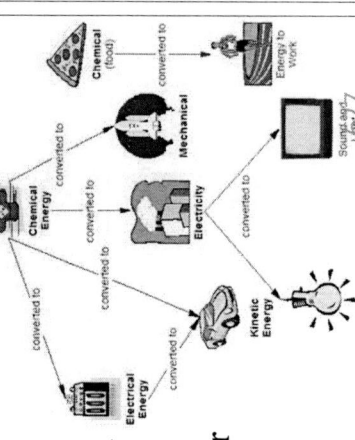

- The 2nd law of thermodynamics states that every exchange of energy increases the entropy of the universe

- Entropy is the disorder, randomness or chaos of molecules in a system undergoing change.

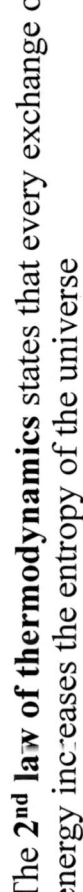

- In an ecosystem, energy conversions are not completely efficient, and some energy is always lost as heat

Law of conservation of mass

- The **law of conservation of mass** states that matter cannot be created or destroyed

- Chemical elements are continually recycled within ecosystems

- In a forest ecosystem, most nutrients enter as dust or solutes in rain and are carried away in water

- Ecosystems are open systems, absorbing energy and mass and releasing heat and waste products

Law of Conservation of Mass Lavoisier (1743-1794)

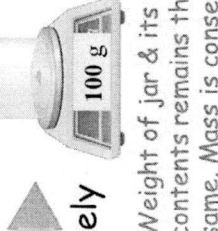

Coal is completely burned

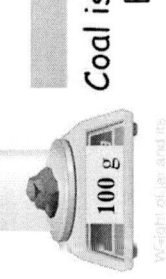

Weight of jar & its contents remains the same. Mass is conserved

In a chemical reaction, matter is neither created nor destroyed, it is transformed into something else.

Producers, Consumers & Decomposers

- Energy and nutrients pass from primary **producers** (autotrophs) to primary **consumers** (herbivores).

- From primary consumers to secondary consumers (carnivores) & then to tertiary consumers.

- **Detritivores**, or decomposers, derive their energy from detritus, nonliving organic matter & help recycle nutrients back to soil.

- Prokaryotes and fungi are important detritivores. Decomposition connects all trophic levels

Primary Production

- Primary production in an ecosystem is the amount of light energy converted to chemical energy by autotrophs during a given time period

- The amount of solar radiation reaching the Earth's surface limits photosynthetic output of ecosystems

- Only a fraction of solar energy is accessible to photosynthetic organisms, and even less is of a usable wavelength

Primary Production in Aquatic Ecosystems

- In marine and freshwater ecosystems, both light and nutrients control primary production

1. Depth of **light penetration** affects primary production in an ocean or lake

2. **Lack of nutrients** limits primary production in ocean & lakes. These are nutrients that are needed for continued growth but are unavailable at times due to limited quantity. Two such elements are Nitrogen and phosphorous.

Primary Production in Terrestrial Ecosystems

- In terrestrial ecosystems, temperature and moisture affect primary production on a large scale along with soil nutrients

- Evapotranspiration can represent the contrast between wet and dry climates

- **Evapotranspiration** is Evaporation + Transpiration.
- Evaporation is loss of water from water bodies & transpiration is loss of water from plants via stomata.
- Evapotranspiration is related to net primary production

Energy transfer between trophic levels is only 10% efficient

- **Secondary production** of an ecosystem is the amount of chemical energy in food converted to new biomass.

- When a caterpillar feeds on a leaf, only about one-sixth of the leaf's energy is used for secondary production

- **Trophic efficiency** is the percentage of production transferred from one trophic level to the next and ranges from 5% to 20%.

- Approximately 0.1% of chemical energy fixed by photosynthesis reaches a tertiary consumer

Production of biomass (g/m²/year)

- Tertiary Consumers: 1
- Secondary Consumers: 10
- Primary Consumers: 100
- Primary Producers: 1000

Nutrients recycling via Biogeochemical cycles

- The biogeochemical cycles involve the biological, geological and chemical factors all in one.

- Life depends on recycling chemical elements

- In a biogeochemical cycle, chemical substances moves through both the biotic and abiotic components of earth

- Carbon, oxygen, sulfur, and nitrogen cycle globally and the less mobile elements phosphorus, potassium, and calcium cycle on a more local level

Many Factors Combine to Affect Biogeochemical Cycles

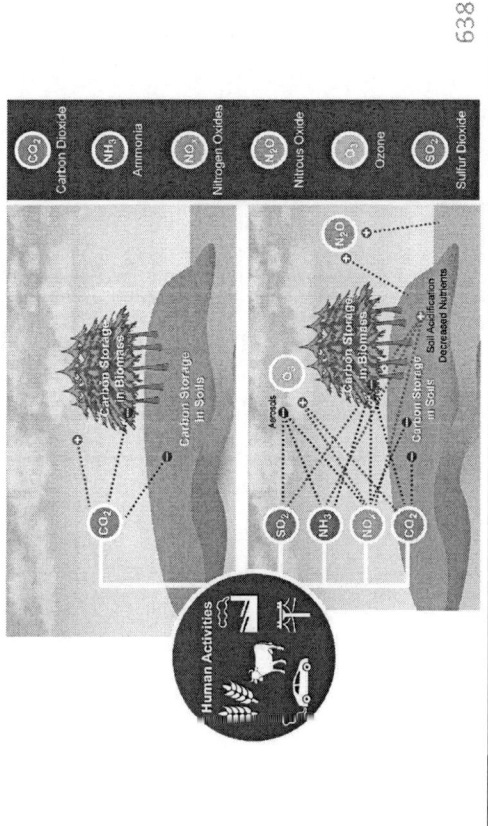

The Carbon Cycle

- Carbon-based organic molecules are essential to all organisms

- Carbon reservoirs include fossil fuels, soils and sediments, solutes in oceans, plant and animal biomass, and the atmosphere

- CO_2 is taken up and released through photosynthesis and respiration; additionally, volcanoes and the burning of fossil fuels contribute CO_2 to the atmosphere

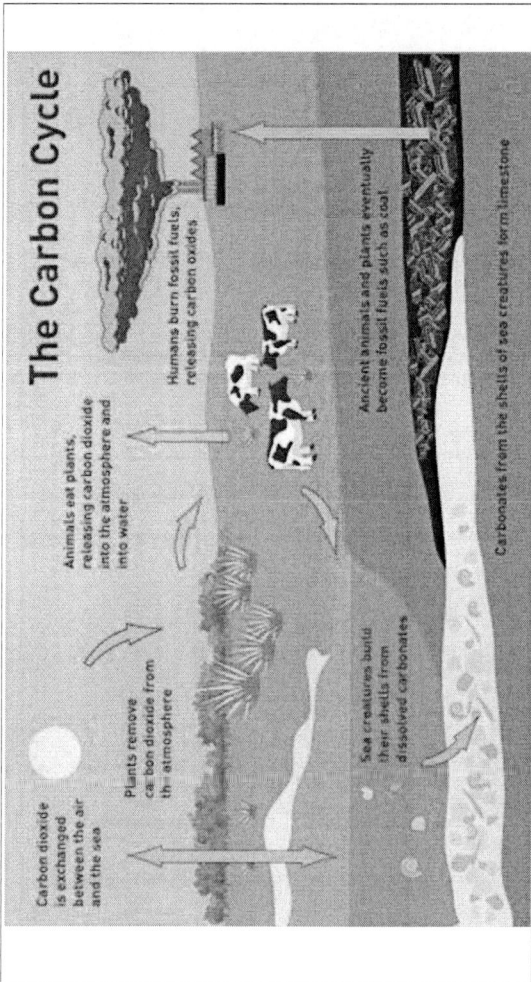

The Nitrogen Cycle

- Nitrogen is a component of amino acids, proteins & nucleic acids

- The main reservoir of nitrogen is the atmosphere (N_2), though this nitrogen must be converted to NH_4^+ or NO_3^- for uptake by plants, since Nitrogen in N_2 form is unusable by plants. 3 steps:

1. Organic nitrogen is decomposed to NH_4^+ by **ammonification**
2. NH_4^+ is decomposed to NO_3^- by **nitrification**
3. NO_3^- back to N_2 by **Denitrification**

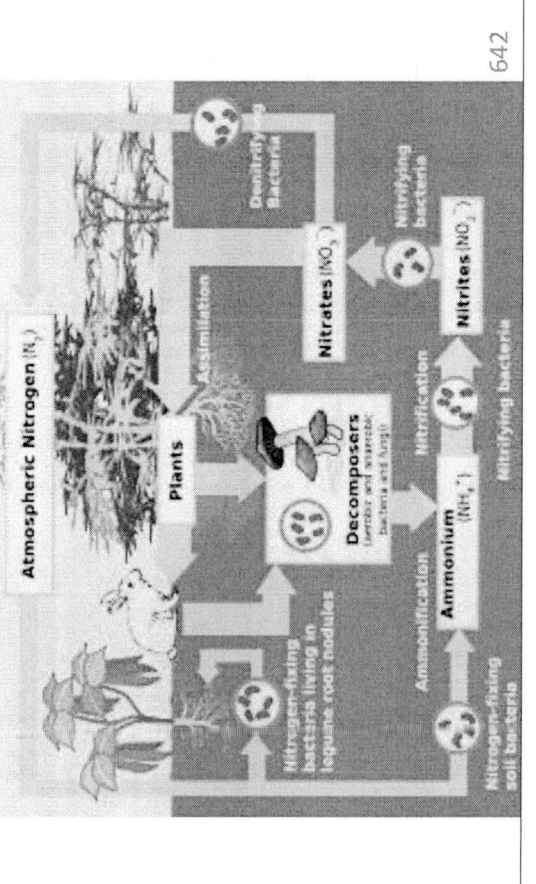

Decomposition and Nutrient Cycling Rates

- Decomposers (detritivores) play a key role in the chemical cycling

- Rates at which nutrients recycle in different ecosystems vary mostly as a result of differing rates of decomposition by due to different environmental factors.

- The rate of decomposition is controlled by temperature, moisture, and nutrient availability

Human Disturbances

- With growth in human population, our activities have disrupted the trophic structure & energy flow of many ecosystems.

- In addition humans have added toxins to the ecosystems

- Agriculture removes from ecosystems nutrients that would ordinarily be cycled back into the soil

- Nitrogen is the main nutrient lost through agriculture; thus, Industrially produced fertilizer is typically used to replace lost nitrogen, but effects on an ecosystem can be harmful

Eutrophication

- **Critical load** for a nutrient is the amount that plants can absorb without damaging the ecosystem

- When excess nutrients are added to an ecosystem, the critical load is exceeded

- Remaining nutrients can contaminate groundwater as well as freshwater and marine ecosystems

- Sewage runoff causes **eutrophication**, excessive algal growth that can greatly harm freshwater ecosystems

Eutrophication

- The excessive algal growth takes the oxygen and nutrients away from other animals like fishes.

- Eventually depleting the animals of their resources and causing the death of an ecosystem.

Acid Precipitation (pH less than 5.6)

- Combustion of fossil fuels is the main cause of acid precipitation

- North American and European ecosystems surrounding industrial regions have been damaged by rain and snow.

- Acid precipitation changes soil pH and causes leaching of calcium and other nutrients.

- Acid precipitation causes damage to coral reefs, kills insects & other aquatic life. It also causes corrosion of steel & skin cancer

- Acid precipitation is the result of gases such as Sulphur dioxide and Nitric oxide reacting with water and forming toxic nitric acid and sulfuric acid.

- These toxic gases released into the atmosphere, return back as acid precipitation

- Environmental regulations have reduced sulfur dioxide emissions but volcanic eruptions continue to be a significant contributor.

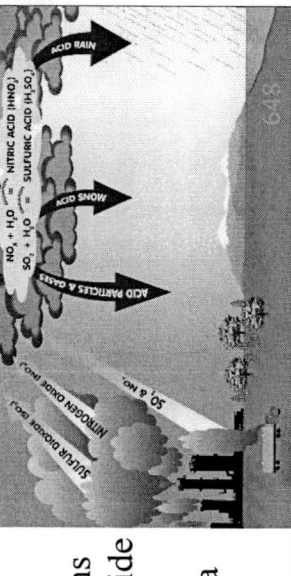

Biological Magnification

- Humans release many toxic chemicals, including synthetics previously unknown to nature

- One reason toxins are harmful is that they become more concentrated in successive trophic levels

- **Biological magnification** concentrates toxins at higher trophic levels, where biomass is lower

- PCBs (polychlorinated biphenyls) and many pesticides such as DDT are subject to biological magnification in ecosystems

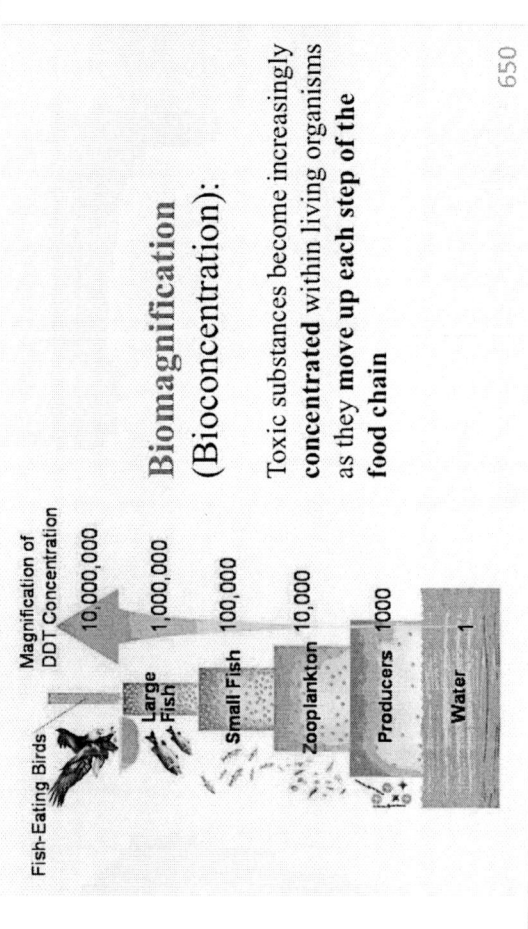

Biomagnification (Bioconcentration):

Toxic substances become increasingly **concentrated** within living organisms as they move up each step of the food chain

Global Warming

- **The Greenhouse effect :** When excess of CO_2, Methane, water vapor, and other greenhouse gases in the atmosphere reflect the infrared radiation back to Earth.

- This has led to warming up of earth and melting of polar ice caps and has been known as global warming. (a very pressing problem of recent times)

- Reflection of infrared radiations in moderate amount is needed to keep the earth's surface warm at a habitable temperature

Greenhouse effect leading to Global Warming

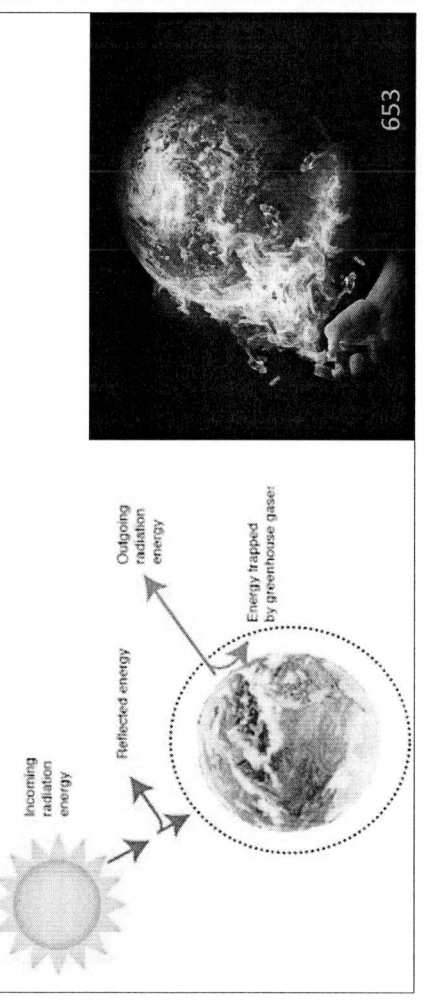

- But Increased levels of these gases are magnifying the greenhouse effect, which causes global warming and erratic climatic change
- Due to the burning of fossil fuels and other human activities, the concentration of atmospheric CO_2 has been steadily increasing
- Global warming can be slowed by reducing energy needs and converting to renewable sources of energy

Depletion of Atmospheric Ozone

- Life on Earth is protected from damaging effects of UV radiation by a protective layer of ozone molecules in the atmosphere
- Destruction of atmospheric ozone results from chlorine-releasing pollutants such as CFCs (chlorofluorocarbons)
- CFC's are volatile derivative of methane, ethane, and propane
- Ozone depletion can causes DNA damage resulting in formation of a dangerous and lethal Cancer.

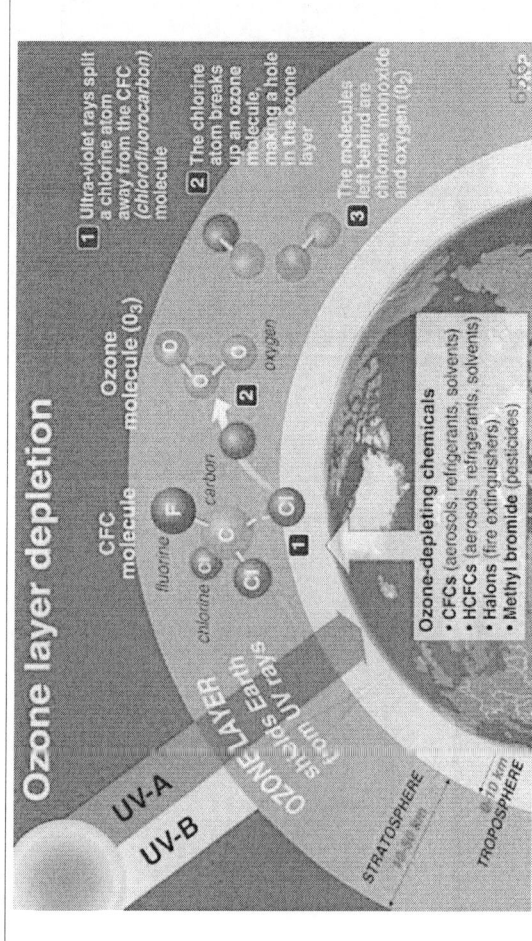

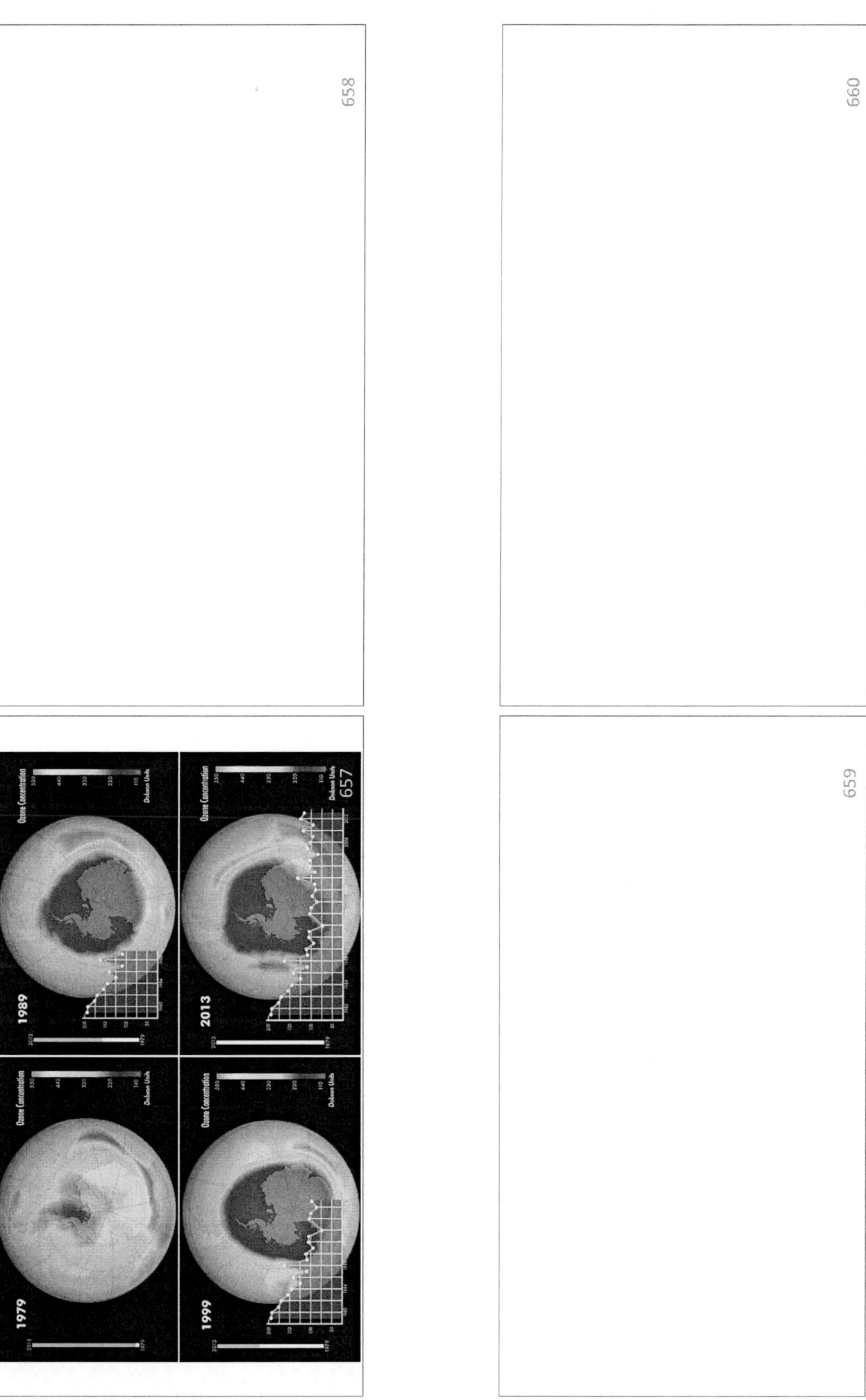

21, Prokaryotes

Prokaryotes (a history of 3.5 billion years)

- **Prokaryotes** are unicellular organisms that lacks a nucleus and membrane-bound organelles.
- Prokaryotes thrive everywhere, including harsh environments that are too acidic, salty, hot or cold for other organisms
- Prokaryotes are microscopic & range from 0.1 to 5.0 μm in diameter compared to eukaryotic cells that range from 10 to 100 μm. But what they lack in structure they make up for in numbers
- Prokaryotes are divided into two domains: **bacteria & archaea**

Interesting Facts about Archaea

- Some Archaea survive at temperatures around 185 degree F
- They have genes similar to bacteria and use operons but their protein synthesis is more similar to eukaryotes
- Archaea live in extreme environments & are called **extremophiles**
    1. Halophiles
    2. Thermophiles
    3. Methanogens

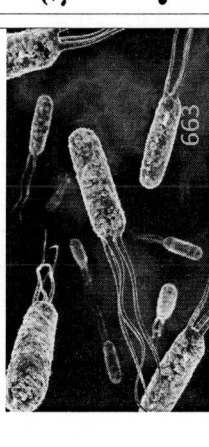

Types of Archaea

1. **Halophiles** live in highly saline environments.
2. **Thermophiles** thrive in very hot environments
3. **Methanogens** live in swamps and marshes and produce methane as a waste product

- Methanogens are strict anaerobes and are poisoned by O_2

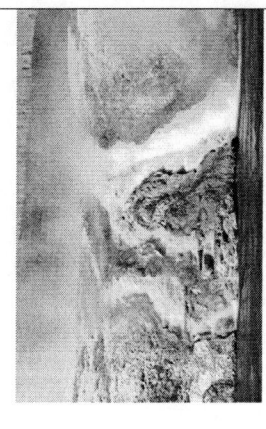

Interesting Facts about Bacteria

- Each one of us is home to 100 trillion bacterial cells which is almost 3 times the no. of human cells in our bodies (37 trillion).

- Prokaryotes replicate and divide (from 1 cell to 2) in 20-40 min, whereas in eukaryotic cell the same process takes 18-24 hours.

- Bacteria have 0.001 times as much DNA as a eukaryotic cell.

- Despite all the difference among genes & proteins between us & bacteria, we both still hold the same 4 bases of DNA (A, T, G C)

Types of Bacteria based on composition cell wall

- Bacteria are divided into two major groups: **gram-positive** and **gram-negative**, based on their reaction to gram staining due to differing cell wall composition.

- Antibiotics target peptidoglycan & damage bacterial cell walls.

- **Gram-negative** bacteria have less peptidoglycan and hence are more likely to be antibiotic resistant than gram positive bacteria.

- Archaea contain polysaccharides and proteins but lack peptidoglycan

Bacteria Cell wall & Antibiotic resistance

- Most bacteria have a cell wall which is made up of **peptidoglycan** (carbohydrates + proteins).

- **Cell wall** gives protection & helps in maintenance of cell shape.

- Their cell wall is the target of our antibiotics since it is toughest layer to attack & penetrate due to the presence of peptidoglycan.

- Unfortunately, the overuse of Antibiotics over the years has selected & promoted certain strains of bacteria that are now antibiotic-resistant.

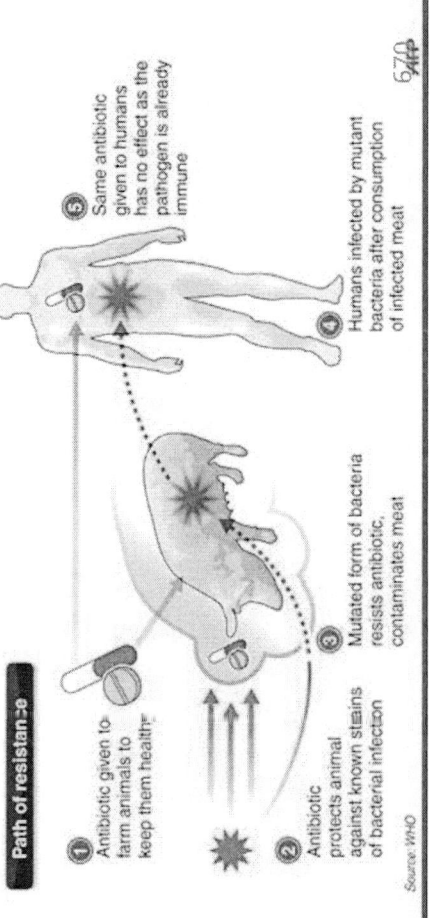

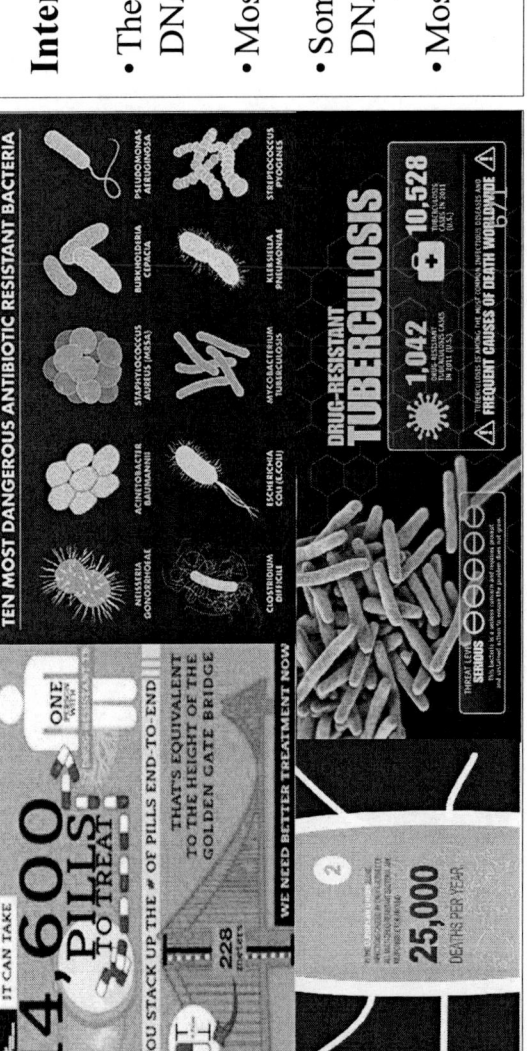

Drugs, animals and mutations

The World Health Organization has detected increased bacterial resistance to antibiotics as a direct result of overuse of drugs in agriculture

Path of resistance

1. Antibiotic given to farm animals to keep them healthy
2. Antibiotic protects animal against known strains of bacterial infection
3. Mutated form of bacteria resists antibiotic, contaminates meat
4. Humans infected by mutant bacteria after consumption of infected meat
5. Same antibiotic given to humans has no effect as the pathogen is already immune

Source: WHO

Internal structures in a Bacterial cell

- The prokaryotic genome is much less than the eukaryotic DNA & is located in a region called **nucleoid**

- Most of the genome consists of a circular chromosome

- Some species of bacteria have additional smaller rings of DNA called **plasmids**

- Most motile bacteria use **flagella** for locomotion (movement)

- Many bacteria exhibit **taxis**, the ability to move towards or away from certain stimulus due to innate behavioral responses
- If the stimulus is a chemical, then the taxis is, chemotaxis and is exhibited by many bacteria including E. coli
- **Chemotaxis**, is a great way to avoid unfavorable conditions and seek optimum surroundings, especially during harsh environments.

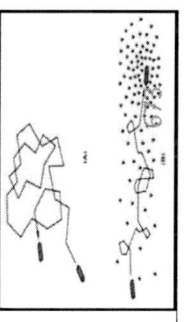

Structural components of a Prokaryotic cell

- Prokaryotic cells have 4 basic structures that are also found in all other life forms, like Plants, Animals, Fungi and Protists.
- These 4 basic structures are DNA, Plasma membrane, Cytosol & Ribosomes.
- Some prokaryotes have **fimbriae** (*attachment pili*), which allow them to stick to a substrate or other individuals in a colony
- **Sex pili** are longer than fimbriae & allow prokaryotes to exchange DNA

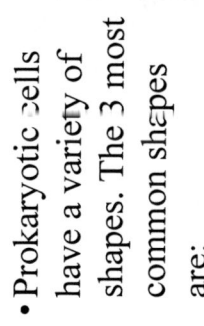

Types of prokaryotes based on cell shapes

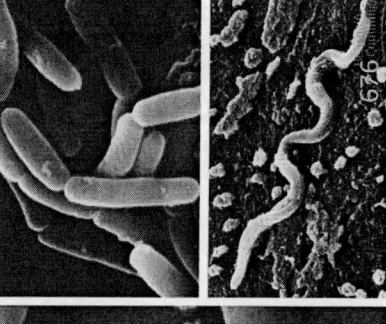
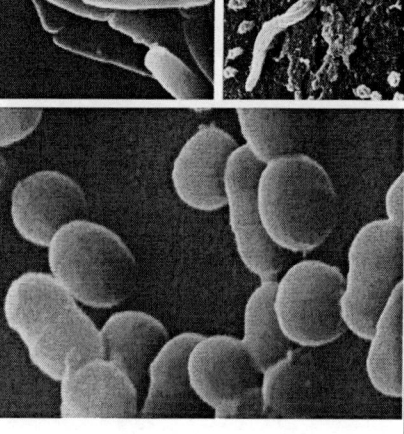

- Prokaryotic cells have a variety of shapes. The 3 most common shapes are:
  1. Spheres (cocci)
  2. Rods (bacilli)
  3. Spirals (spirilla)

Types of prokaryotes based on Nutrition

- **Chemoautotrophs**: obtain energy from inorganic chemicals and carbon from CO_2
- **Photoautotrophs**: obtain energy from sunlight and carbon from CO_2
- **Photoheterotrophs**: obtain energy from sunlight and carbon from Organic compounds
- **Chemoheterotrophs**: obtain energy from organic compounds and carbon from Organic compounds as well

The Prokaryotic genius- Endospores

- Many prokaryotes form metabolically inactive **endospores**, which can remain viable in harsh conditions for centuries.

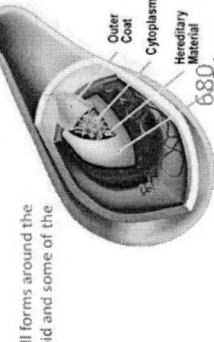

If environmental conditions become unfavourable for growth (such as lack of food), a bacterium may produce an endospore.

A tough shell forms around the cell's nucleoid and some of the cytoplasm.

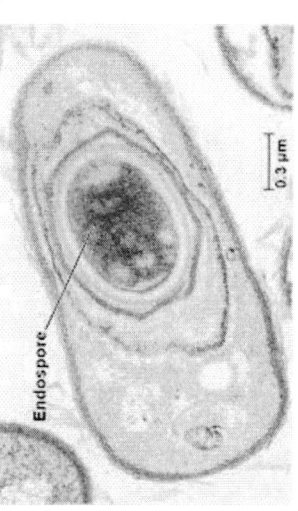

Types of prokaryotes based on metabolism in respect to O_2

1. **Obligate aerobes** require O_2 for cellular respiration
2. **Obligate anaerobes** are poisoned by O_2 & use fermentation or anaerobic respiration
3. **Facultative anaerobes** can survive with or without O_2

(a) Obligate aerobes (b) Obligate anaerobes (c) Facultative anaerobes (d) Aerotolerant anaerobes

Replication & Mutation- tools of prokaryotic success

- Prokaryotes have a very short generation time and reproduce by Binary fission in 20- 40 min
- High frequency of Replication makes them more prone to genetic mutations introduced by random errors.
- This high frequency of mutation in turn endows Prokaryotes with an astonishing genetic diversity
- They were the 1st ones to inhabit the Universe and in all likelihood will be last ones of those who survive.

Another Prokaryotic genius Metabolic cooperation

- Another significant adaptation seen in prokaryotes is the mutual metabolic cooperation that exists between them.

- This cooperation allows them to use environmental resources they would not be able to use as individual cells. For e.g.:

- In the cyanobacterium *Anabaena*, photosynthetic cells & nitrogen-fixing cells called **heterocytes** exchange metabolic products.

Diagram on next slide→

Metabolic cooperation

- Metabolic cooperation in filamentous growth colonies (as in *Anabaena*)

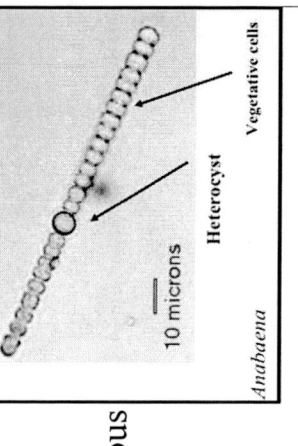
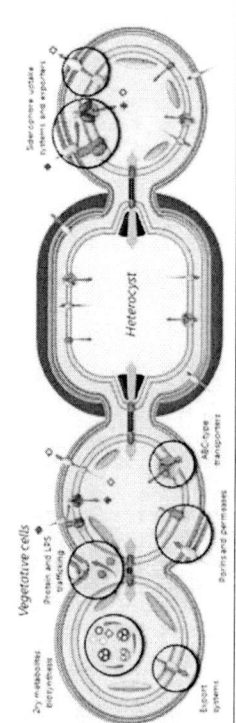

Metabolic cooperation

- Metabolic cooperation in surface-coating colonies called **biofilms**

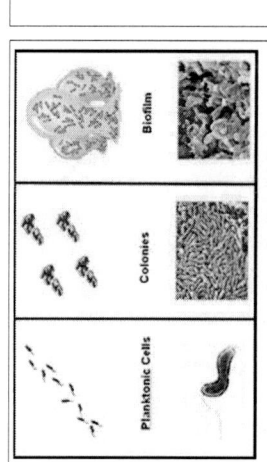
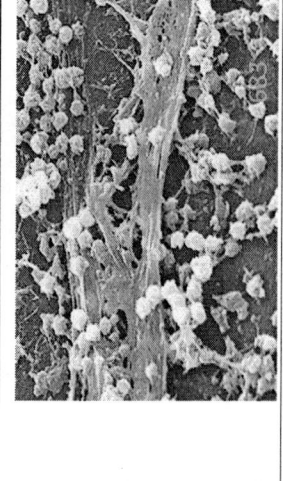

Prokaryotes – Masters of Adaptation

- Prokaryotes evolve rapidly because of their short generation times

- Prokaryotic cells are much simpler and lack compartmentalization that's seen in a Eukaryotic cell

- Characteristics such as Chemotaxis, Endospores, mutual cooperation & having a short generation time (with frequent mutations), have turned them into masters of Adaptation.

Factors that promote genetic diversity in Prokaryotes

- Prokaryotes have huge genetic variation and the three factors that contribute to this diversity are-

1. **Rapid reproduction**- Prokaryotes reproduce by binary fission.

2. **Mutation**- Mutation rates during binary fission are low, but due to rapid reproduction, mutations accumulate very rapidly.

3. **Genetic recombination** - Additional diversity arises from genetic recombination.

Genetic Recombination

- Genetic recombination – in prokaryotes can happen in one of the 3 ways, transformation, transduction, and conjugation

1. **Transformation:** A prokaryotic cell can taking up and incorporate foreign DNA from the surrounding environment

2. **Transduction :** The movement of genes between bacteria via bacteriophages (viruses that infect bacteria)

3. **Conjugation :** The process of transferring genetic material is between bacterial cells via sex pili

Conjugation via Sex pili

- Sex pili allow cells to connect and come close together for a DNA transfer to take place

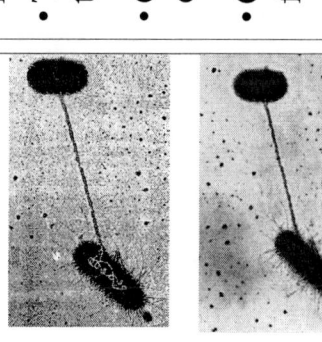
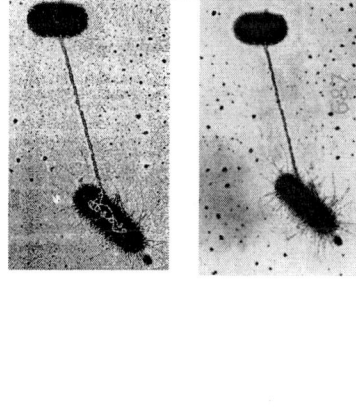

The F Factor in a Plasmid

- A piece of DNA called the **F factor** is required for the production of sex pili & is transferable during conjugation

- The F factor can exist as a separate plasmid or as DNA within the bacterial chromosome.

- Cells containing the **F plasmid** function as DNA donors during conjugation.

- Cells without the F factor function as DNA recipients

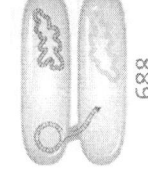
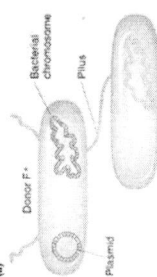

The F Factor in the Chromosome

- A cell with the F factor built into its chromosomes functions as a donor during conjugation.

- The recipient becomes a recombinant bacterium, with DNA from two different cells

R Plasmids and Antibiotic Resistance

- **R plasmids** carry genes for antibiotic resistance

- Antibiotics have been selecting for bacteria with genes that are resistant to antibiotics

- This has led to an overabundance of the antibiotic resistant strains of bacteria and it now poses a great danger to all life forms

When not to use Antibiotics

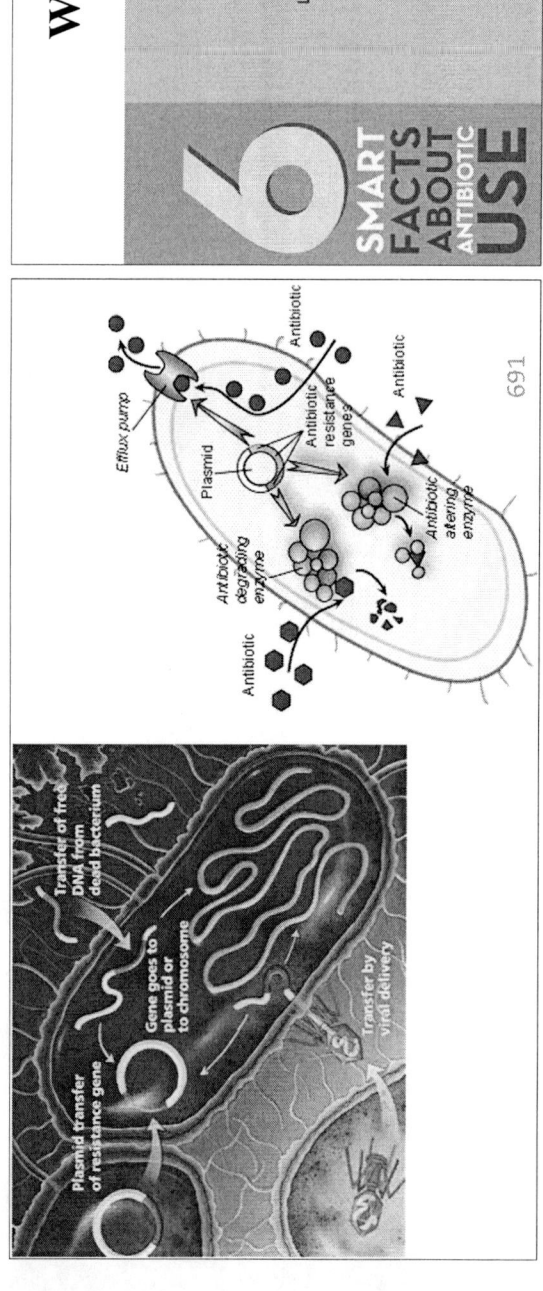

Prokaryotes crucial role in Chemical cycling

- Prokaryotes play a major role in the recycling of chemical elements (nutrients) between the living and nonliving components of ecosystems

- Chemoheterotrophic prokaryotes function as **decomposers**, breaking down corpses, dead vegetation, and waste products

- By doing so, they return all the organics and inorganic nutrients back to the soil for the uptake by plants.

Decomposers

By assisting in breaking down dead organisms, prokaryotes supply raw materials in the environment.

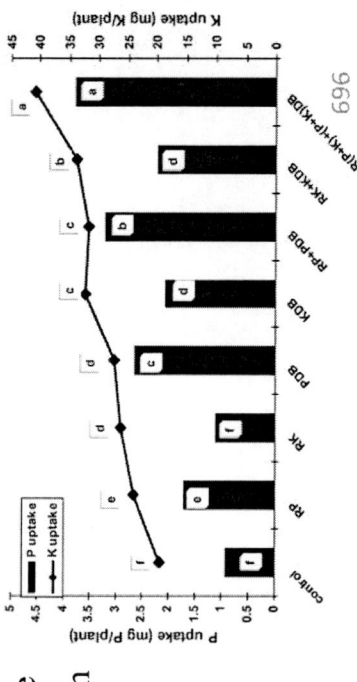

Bacteria of the genus *Rhizobium*

Cyanobacteria of the genus *Anabaena*

Bacteria called actinomycetes

- Nitrogen-fixing prokaryotes add usable nitrogen to the environment

- Prokaryotes can also increase the availability of phosphorus, and potassium for plant growth

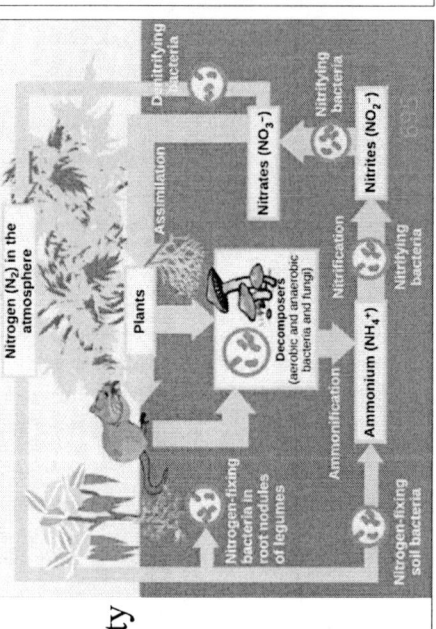

Phosphorus (**P**) uptake and potassium (**K**) uptake (mg/plant) of maize plant as affected by rock minerals and inoculation with **P** and **K**-dissolving bacteria.

RP=rock phosphate
RK=rock potassium
PDB=phosphate dissolving bacteria
KDB=potassium dissolving bacteria

Ecological Interactions - of 3 kinds

- **Symbiosis** is an ecological relationship in which two species are intimate with one another (live in close contact).
- Symbiosis exists between a larger **host** & a smaller **symbiont**
- Prokaryotes form symbiotic relationships with larger organisms. The 3 kinds of Symbiosis are:
  1. Commensalism
  2. Parasitism
  3. Mutualism

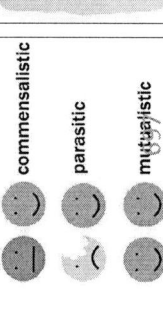

Mutualism | Parasitism | Commensalism

- In **Mutualism**, both symbiotic organisms benefit
- In **Parasitism** an organism called a **parasite** harms but does not kill its host
- In **commensalism**, one organism benefits while other remains unaffected

Commensal Bacteria
Microorganisms that populate the gastrointestinal tract

More on Parasitism

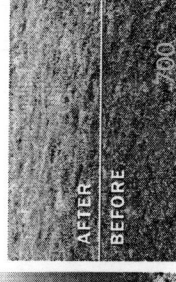

- The disease causing Parasites are called Pathogens.
- Pathogenic prokaryotes typically cause disease by releasing exotoxins or endotoxins
- **Exotoxins** are toxic substances secreted by bacteria & released outside the cell. They cause disease even if the prokaryotes that produce them are not physically present.
- **Endotoxins** are toxins located within the cell of a bacteria & are released when bacteria die & their cell walls break down

The Positive role of Prokaryotes

- Prokaryotes play a very critical role in **bioremediation**, the use of organisms to remove pollutants from the environment
- Bioremediation helps to clean soil & groundwater by microbes that use the toxic pollutants as source of food & energy

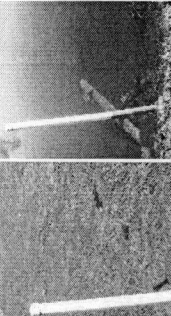
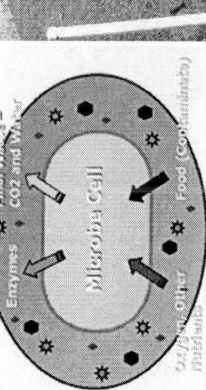